Springer Theses

Recognizing Outstanding Ph.D. Research

Aims and Scope

The series "Springer Theses" brings together a selection of the very best Ph.D. theses from around the world and across the physical sciences. Nominated and endorsed by two recognized specialists, each published volume has been selected for its scientific excellence and the high impact of its contents for the pertinent field of research. For greater accessibility to non-specialists, the published versions include an extended introduction, as well as a foreword by the student's supervisor explaining the special relevance of the work for the field. As a whole, the series will provide a valuable resource both for newcomers to the research fields described, and for other scientists seeking detailed background information on special questions. Finally, it provides an accredited documentation of the valuable contributions made by today's younger generation of scientists.

Theses are accepted into the series by invited nomination only and must fulfill all of the following criteria

- They must be written in good English.
- The topic should fall within the confines of Chemistry, Physics, Earth Sciences, Engineering and related interdisciplinary fields such as Materials, Nanoscience, Chemical Engineering, Complex Systems and Biophysics.
- The work reported in the thesis must represent a significant scientific advance.
- If the thesis includes previously published material, permission to reproduce this must be gained from the respective copyright holder.
- They must have been examined and passed during the 12 months prior to nomination.
- Each thesis should include a foreword by the supervisor outlining the significance of its content.
- The theses should have a clearly defined structure including an introduction accessible to scientists not expert in that particular field.

More information about this series at http://www.springer.com/series/8790

Maosheng Zheng • Jinren Ni

Nitrogen Removal Characteristics of Aerobic Denitrifying Bacteria and Their Applications in Nitrogen Oxides Emission Mitigation

Doctoral Thesis accepted by
Peking Univerisity, Beijing, China

Springer

Author
Dr. Maosheng Zheng
College of Environmental Science
and Engineering, MOE Key Laboratory
of Resources and Environmental Systems
Optimization
North China Electric Power University
Beijing, China

Supervisor
Prof. Jinren Ni
College of Environmental Sciences
and Engineering, MOE Key
Laboratory of Water and Sediment
Sciences, Ministry of Education
Peking University
Beijing, China

ISSN 2190-5053 ISSN 2190-5061 (electronic)
Springer Theses
ISBN 978-981-13-2431-4 ISBN 978-981-13-2432-1 (eBook)
https://doi.org/10.1007/978-981-13-2432-1

Library of Congress Control Number: 2018956284

This Springer imprint is published by the registered company Springer Nature Singapore Pte Ltd.
The registered company address is: 152 Beach Road, #21-01/04 Gateway East, Singapore 189721, Singapore

Supervisor's Foreword

Aerobic denitrification has received extensive attentions in recent years as an efficient biological nitrogen removal technology. However, most aerobic denitrifying bacteria produced N_2O instead of N_2 as the final denitrification product, which impaired their potential applications in wastewater treatment. In this thesis study, Dr. Maosheng Zheng has systematically investigated the nitrogen removal characteristics of two screened aerobic denitrifying bacteria and their applications in nitrogen oxides emission reduction. It was well demonstrated that strain *Pseudomonas stutzeri* PCN-1 possessed excellent capacity of aerobic nitrogen removal whether nitrate, nitrite, or N_2O were taken as denitrification substrates. RT-PCR analysis revealed that the coordinate expression of denitrification genes was the fundamental reason for rapid N_2O reduction. Bioaugmentation experiments conducted in denitrifying SBR and pilot-scale Carrousel oxidation ditch confirmed the strain could significantly enhance denitrification performance, reduce N_2O emission and improve system stability. Strain *P. aeruginosa* PCN-2 accumulated negligible NO during aerobic nitrate and nitrite removal. A biotrickling filter establish with the strain achieved NO removal efficiency of 91.94–96.74% from flue gas at inlet O_2 concentration 0–10% and NO concentration 100–500 ppm. This study is of great significance in potential application of aerobic denitrification in mitigating nitrogen oxides emission from biological nitrogen removal system.

Beijing
June 2018

Jinren Ni

Parts of This Thesis Have Been Published in the Following Journal Articles

Zheng M., Li C., Liu S., Gui M., Ni J. 2016. Potential application of aerobic denitrifying bacterium Pseudomonas aeruginosa PCN-2 in nitrogen oxides (NOx) removal from flue gas. *Journal of Hazardous Materials*, 318, 571–578.

Zheng M.S., He D., Ma T., Chen Q., Liu S.T., Ahmad M., Gui M.Y., Ni J.R.* 2014. Reducing NO and N_2O emission during aerobic denitrification by newly isolated *Pseudomonas stutzeri* PCN-1. Bioresource Technology 162, 80–88.

Zheng M.S., Tian Y.H., Liu T., Ma T., Li L., Li C., Ahmad M., Chen Q., Ni J.R.* 2015. Minimization of nitrous oxide emission in a pilot-scale oxidation ditch: generation, spatial variation and microbial interpretation. Bioresource Technology 179, 510–517.

Acknowledgments

Five years of doctoral time flashed by, and twenty years of student career would soon come to an end. In the most important five years of my life, there have been confusions and hesitations, pessimisms and losses, but more was the struggle, harvest, and growth. Upon the completion of this thesis, I would like to express my heartfelt thanks to all my teachers and friends for their guidance and care.

First of all, I would thank my supervisor, Professor Ni Jinren. This thesis was completed under his professional and patient guidance. In particular, his passion in academic research and profound attainments impressed and benefited me greatly and would continue to encourage and guide me to be a potential scientist.

Second, I would like to extend my sincere thanks to Professor Meiping Tong, Zhengfang Ye, Tianhong Li, Weiling Sun, Sitong Liu, Huazhang Zhao, and Chen Qian from the Department of Environmental Engineering. Their suggestions provide a great help for me to finish my doctoral research. Thank Xiaoping Ding, Lifang Ye, Guiping Zhu, and Fenyong Zhao from Langfang laboratory for giving the enthusiastic experimental help. Thank Feifei Zhang, Dongkang Fu, Baogang Zhang, Shuo Yao, Xuan Xing, Lina Pang, and Li Lei, who have given me a lot of guidance and encouragement to avoid detours. I would like to thank Tao Ma, Tang Liu, Yuhao Tian, Shuangyue Zhu, Li Li, Mengyao Gui, Chenyuan Dang, and Shufeng Liu in the laboratory for the help and support.

Third, I would express my thanks to my parents who give me the life and healthy body, the spiritual support and the encouragement. I hope the progress achieved in my long school career can give them some comfort.

At last, I would thank my girlfriend Shuo Xu for the understanding and support to my research. I spent one third of my life full of colorful happiness because of her company. I believe all our good wishes will come true in the near future, just like the appointment to go to Peking University together made by the two ingenuous undergraduate students that year.

I dedicate this to all the teachers, students, relatives, and friends who care and support me!

Beijing
April 2015

Maosheng Zheng

Abstract

With the deeper understanding of nitrogen pollution in water, stricter wastewater discharge standards were enacted, which put forward a higher requirement on biological nitrogen removal (BNR) technologies. Aerobic denitrification was a new type of BNR technology, which has achieved rapid development in recent years due to its advantages of simple process, high efficiency, and no needing to add acid or alkali. However, most aerobic denitrifying bacteria produced N_2O rather than N_2 as the final denitrification product due to the sensitivity of N_2O reductase to O_2, which puts forward new challenges in the control of greenhouse gas emission like N_2O from wastewater treatment process.

In this study, the characteristics of aerobic denitrification and nitrogen oxides reduction were systematically investigated based on two screened aerobic denitrifying bacterial strains, *Pseudomonas stutzeri* PCN-1 and *Pseudomonas aeruginosa* PCN-2.

Strain *P. stutzeri* PCN-1 possessed excellent capacity of aerobic denitrification and N_2O reduction. Nitrogen removal rates of 11.66, 12.80, and 9.67 mg L^{-1} h^{-1} were achieved by strain PCN-1 when nitrate, nitrite, and N_2O were taken as denitrification substrates, respectively. Moreover, the maximal NO and N_2O accumulation were as low as 0.003% and 0.33% of removed NO_3-N or 0.006% and 0.29% of reduced NO_2-N, demonstrating the strain could perform rapid N_2O reduction to N_2 during aerobic denitrification. Reverse transcription quantitative PCR analysis revealed that the coordinate expression of denitrification genes *nirS*, *cnorB*, and *nosZ* was the fundamental reason for low NO and N_2O accumulation. The optimal conditions for the highest denitrifying rate and lowest NO and N_2O accumulation were achieved with succinate or acetate as carbon source, pH 7–9, and C/N 4–6. The normal denitrification and N_2O reduction under 0–100% O_2 indicated the strain's strong tolerance to high O_2 concentrations.

Strain *P. stutzeri* PCN-1 could significantly enhance denitrification performance, reduce NO and N_2O emission, and improve stability of activated sludge systems. First, when it was inoculated into activated sludge in shaking flask, aerobic nitrogen removal performances of the system were significantly improved and NO and N_2O

emission were significantly reduced with ammonia, nitrite, or nitrate as main pollutants. Moreover, when strain PCN-1 was inoculated into long-term denitrifying SBR, N_2O emissions were reduced by 19.81%, 46.12%, and 45.13% at aerated O_2 concentration of 0%, 5%, and 10%, respectively. Finally, strain PCN-1 was applied in a pilot-scale Carrousel oxidation ditch system, and it was found that N_2O and NO emission were reduced by 36.6% and 23.0% when the system was steadily operated at SRT of 17 days, implying the strain's great potential applications in nitrogen oxides reduction in activated sludge system. In addition, N_2O and NO emission amount was reduced by 49.13% and 69.87% when the system suffered from a one-hour "high ammonia shock" with four-time normal ammonia load and reduced by 59.7% and 64.0% when the system suffered from a two-hour "aeration failure shock" with a sudden reduction of aeration rate by 50%, demonstrating that the strain PCN-1 played an important role in maintaining steady operation of activated sludge system.

Strain *P. aeruginosa* PCN-2 performed excellent capability of aerobic denitrification and nitrogen oxides reduction. Negligible NO was accumulated during aerobic denitrification with nitrate or nitrite as nitrogen source, which was mainly attributed to the coordinate expressions of genes *napA*, *nirS*, and *cnorB* and high expression amount of gene *cnorB*. Chemical reaction of NO and O_2 also made a great contribution. Standing sealed bottle experiments demonstrated that immobilized strain PCN-2 exhibited significant NO and NO_2 removal capability under O_2 concentration of 0–10%. The strain was finally applied in biotrickling filter to establish a flue gas denitrification system, and NO_x removal efficiency of 91.94–96.74% was achieved at inlet O_2 concentration of 0–10% and NO concentration of 100–500 ppm, which realized high NO_x removal efficiency under aerobic conditions.

This study is of great significance for deep understandings of aerobic denitrification and potential application in the control of nitrogen oxides emission from BNR system.

Keywords Aerobic denitrification, Nitrogen oxides reduction, Bioaugmentation, Activated sludge

Contents

Chapter 1
Introduction

With the acceleration of industrialization and the rapid development of social economy, a large number of industrial and domestic wastewater is produced in China. The discharge of sewage to the receiving water body leads to the deterioration of water environment and poses a great threat to human health. The pollutants in the wastewater mainly include organic matter, heavy metals, and nutrients like nitrogen and phosphorus. Particularly, the presence of excessive nitrogen and phosphorus in water body could induce eutrophication, resulting in serious ecological environmental problems such as algae bloom, blackwater, dead fish, and so on. In order to deal with the increasingly serious water issues, China has enacted more stringent sewage discharge standards. "Urban sewage treatment plant pollutant discharge standards" (GB18918-2002) will be a discharge standard divided into standard A and standard B, with a higher emission requirement on total nitrogen imposed on standard A. The recently launched "Environmental Protection Law of the People's Republic of China (2014 Revision)" on the sewage enterprises and units has put forward stringent water treatment requirements and regulatory measures. Therefore, it has become focus of research in the field of water pollution control to effectively reduce the nitrogen pollution in the water and develop energy-saving nitrogen removal technology.

1.1 Biological Nitrogen Removal (BNR) Theory and Technology

1.1.1 The Traditional BNR Theory and Technology

The traditional biological nitrogen removal theory is the sequential nitrification and denitrification based on the bacterial metabolisms. Firstly, the organic nitrogen in wastewater is decomposed and transformed into ammonia nitrogen by heterotrophic

M. Zheng, *Nitrogen Removal Characteristics of Aerobic Denitrifying Bacteria and Their Applications in Nitrogen Oxides Emission Mitigation*, Springer Theses,
https://doi.org/10.1007/978-981-13-2432-1_1

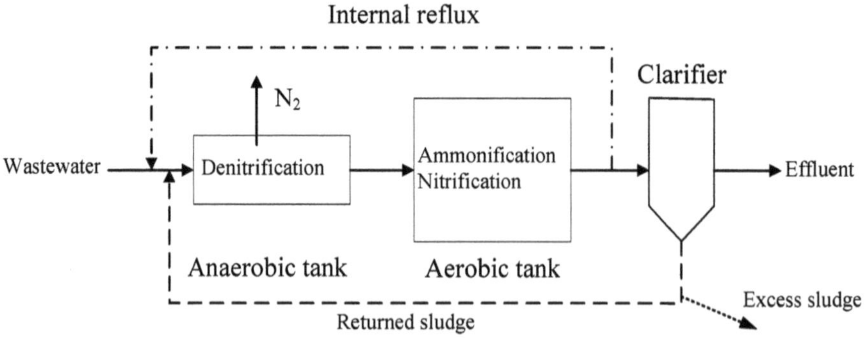

Fig. 1.1 Conventional A/O biological nitrogen removal activated sludge process

bacteria, which is a relatively easier process to carry out. Secondly, ammonia nitrogen was converted to nitrite by the autotrophic ammonia-oxidizing bacteria (AOB) and further transformed into nitrate by nitrite-oxidizing bacteria (NOB). Hereby, the complete process of nitrification is achieved by the joint effort of AOB and NOB. Finally, nitrate in wastewater is converted into gaseous nitrogen by the heterotrophic denitrifying bacteria, which is known as denitrification. Nitrification needs to be carried out under aerobic condition, the general requirement for dissolved oxygen is higher than 2.0 mg L^{-1}, and inorganic carbon as carbon source is needed for bacterial growth and metabolism; denitrifying bacteria need organic carbon for cell growth, and this process can only be carried out under hypoxic conditions. Generally, the dissolved oxygen is below 0.5 mg L^{-1} to guarantee the reducing environment.

Due to the different requirements of nitrifying bacteria and denitrifying bacteria for nutrients and dissolved oxygen, the biological denitrification process based on the traditional biological denitrification theory can only separate the nitrification process and denitrification process, such as A/O, the most typical pre-denitrification process based on activated sludge (Fig. 1.1). However, this separation brings about lots of drawbacks, like increasing infrastructure and operating costs, requiring alkali and carbon additional, etc.

1.1.2 The New BNR Theory and Technology

In recent years, with the in-depth study of microbiology and mastery of biotechnology, bio-denitrification technology has also been further developed from the simple process to the biological characteristics of the process to promote the development direction of reform in order to achieve the high efficiency and low consumption purpose. Some processes break through the traditional biological denitrification theories such as anaerobic ammonium oxidation (ANAMMOX), shortcut nitrification and denitrification, simultaneous nitrification and denitrification, and aerobic denitrification. Aerobic denitrification and other new biological nitrogen removal process more and more attention.

1. Anaerobic Ammonium Oxidation

Anaerobic ammonium oxidation is the biological oxidation process where the microorganisms under the anaerobic condition directly convert ammonia nitrogen (as the electron donor) and nitrite nitrogen (as the electron acceptor) into nitrogen gas. In 1995, Mulder et al. (1995) found ammonia nitrogen losses and nitrogen generation in an anaerobic denitrification fluidized bed reactor. Anaerobic ammonium oxidation was reported for the first time. At present, the researchers have been domesticated to a variety of anaerobic ammonia-oxidizing bacteria that are a class of energy autotrophic bacteria, mainly concentrated in the mold fungus (Schmidt et al. 2003).

Compared with traditional nitrogen removal process, anaerobic ammonium oxidation process can save carbon source, reduce energy consumption, and reduce sludge disposal cost and has obvious advantages in implementation and running cost for wastewater with high ammonia nitrogen and low carbon to nitrogen ratio. However, because the anaerobic ammonium oxidation bacteria proliferation rate is very slow, generation cycle is very long, so the biggest problem lies in the long start-up time. Also because the organic matter in the sewage will lead to heterotrophic bacteria breeding and the autotrophic bacteria in the reactor matrix formation of competitive inhibition resulting in bacterial population distribution, normal metabolism is destroyed. Therefore, how to maintain anaerobic ammonia oxidation bacteria in the stabilization of function in the reactor is another problem to be solved urgently.

2. Shortcut Nitrification and Denitrification

The traditional biological denitrification theory of the entire process of denitrification needs two steps of nitrification and denitrification. Nitrification and denitrification in the whole process refer to the transformation of ammonia nitrogen into nitrite under the action of ammonia-oxidizing bacteria (AOB) and then further transformed to nitrate under the action of nitrite-oxidizing bacteria (NOB). The denitrification process is based on nitrification products of nitrate, nitrite as electron acceptors, with their role in denitrifying bacteria to reduce gaseous products. It can be seen that nitrite is not only an intermediate product of nitrification but also a substrate of denitrification, so if the ammonia nitrogen is oxidized to nitrite without further oxidation to nitrite, then nitrite salt denitrification for the electron acceptor can achieve short-range nitrification and denitrification. Compared with the traditional nitrification and denitrification process, nitrification-denitrification has many advantages: nitrification stage to save 25% oxygen and denitrification stage to save 40% carbon source; it reduces the final sludge production, reduces nitrification process alkali dosage, and reduces the reactor volume by 30–40%.

The theoretical basis for the short-range denitrification is that ammonia-oxidizing bacteria and nitrite-oxidizing bacteria differ in the best survival conditions. Artificial environment is suitable for ammonia-oxidizing bacteria but inhibits the growth of nitrite-oxidizing bacteria. The oxidizing bacteria are enriched in the reactor and occupy a competitive advantage to exclude the nitrite-oxidizing bacteria from the system. Thus, the nitrite nitrogen can be accumulated in the reactor in order to further complete the beginning of nitrite denitrification. SHARON (single reactor for high activity ammonia removal over nitrite) process and OLAND (oxygen-limited autotrophic nitrification-denitrification) process are short-range nitrification and denitrification technologies commonly used in the treatment process.

The key to shortcut nitrification is to control ammonia at the nitrite stage, and nitrite is used as an electron acceptor for denitrification. The factors influencing the accumulation of nitrite are temperature, pH, DO, and free ammonia. When the temperature is higher than 15 °C, the sludge age of ammonia-oxidizing bacteria is shorter than that of nitrite-nitrifying bacteria, and as the temperature increases, the gap becomes larger. So you can use high temperature control sludge age to eliminate the generation of longer nitrifying bacteria, to achieve the accumulation of nitrite. The accumulation of nitrite can also be achieved by the different growth rates of ammonia-oxidizing bacteria and nitrite-oxidizing bacteria at different pH values. The pH value is between 7.5 and 9.0. With the increase of pH, the accumulated concentration of nitrite increases. When the dissolved oxygen was 0.7–1.4 mg L^{-1}, the accumulated concentration of nitrite increases significantly with the oxidation of ammonia, so the dissolved oxygen could be controlled in a lower range to achieve the accumulation of nitrite. The higher concentration of free ammonia has a significant inhibitory effect on the nitrite oxidation, so it is easier to achieve shortcut nitrification and denitrification for high ammonia wastewater treatment.

3. Simultaneous Nitrification and Denitrification

In the traditional nitrification and denitrification theory, the denitrification process has been regarded as a strict anaerobic process. Denitrifying bacteria as facultative bacteria preferentially use O_2 to breathe, even at DO concentration level as low as 0.1 mg L^{-1}, which prevents nitrate and nitrite as electron acceptors. In recent years, however, non-assimilative total nitrogen loss under aerobic conditions has been consistently observed in practice, and even in many practical aerobic nitrification tanks, total nitrogen losses have reached 30% (Daigger and Littleton 2000). The simultaneous nitrification and denitrification phenomena are mainly explained by macroscopic environmental theory, microenvironmental theory, and microbiology theory.

Macroenvironmental theory holds that even in aerobic activated sludge systems, when a point source aeration device or aeration is not uniform, there will still be a wide range of local anoxic environment. For example, in biofilm reactors, anoxic regions are formed in the membrane due to the influence of substrate concentration and film thickness variations. Other phenomena such as SBR reactors and oxidation ditches are similar. In fact, in a production-scale bioreactor, a completely uniform mixing state does not exist, so a certain degree of denitrification, i.e., simultaneous nitrification and denitrification, is entirely possible in the aeration stage.

According to the microenvironmental theory, dissolved oxygen gradient is formed in the microbial flocs or biofilm due to oxygen diffusion limitation, which leads a higher dissolved oxygen concentration outside the surface of microbial flocs or biofilm favoring the growth of aerobic nitrifying bacteria. In the in-depth flocs, hypoxic environment is formed due to oxygen transfer and external oxygen consumption which provide a suitable condition for denitrifying bacteria to conduct nitrate reduction. Thereby simultaneous nitrification and denitrification are achieved in the same sludge flocs.

The microbiological theory suggests that aerobic denitrifying bacteria can simultaneously take oxygen and nitrate or nitrite as electron acceptors and contribute significantly to the total nitrogen removal under aerobic conditions. This theory also overthrows the past theory that denitrifying bacteria are all strict anaerobic type. This part will be discussed in the subsequent chapters.

Simultaneous nitrification and denitrification process make it possible to put the nitrification and denitrification process in one reactor, which shows attractive advantages, such as saving the reactor volume and structure area, reducing the cost of infrastructure. Alkalinity dosing can also be reduced as alkalinity produced in the denitrification reaction can compensate for that consumed by nitrification. Also SND can promote partial nitrification and denitrification, avoiding nitrite oxidation to nitrate, and then reduced back to nitrite, which shorten reaction time and save aeration and carbon source consumption.

1.2 Research Progress on Aerobic Denitrification

1.2.1 Aerobic Denitrifying Bacteria and Characteristics

The conventional denitrification theory suggested that the strong electron-accepting ability of oxygen could prevent the electron transfer to the denitrifying substrate nitrite or nitrate and thus inhibit the denitrification process. In the 1980s, aerobic denitrifying bacteria were first isolated in the sulfur removal and denitrification systems (Robertson and Kuenen 1984), named as *Thiosphaera pantotropha* (now renamed *Paracoccus denitrificans*). The strain is spherical or short rodlike, arranged in pairs or in columns, where the optimum growth temperature is 37 °C, and it can grow in the range of 15–42 °C with the optimum pH range of 6.5–10.5. It can still reduce nitrate in the water dissolved oxygen saturation of 90%. Therefore the discovery of aerobic denitrifying bacteria provides a reliable way to achieve simultaneous nitrification and denitrification in a single reactor. This characteristic has attracted wide attentions, and more aerobic denitrifying bacteria were isolated from several genera, including *Alcaligenes*, *Pseudomonas*, *Achromobacter*, *Acinetobacter*, *Comamonas*, *Bacillus*, etc. (Chen and Ni 2011; Chen et al. 2015; Huang et al. 2013; Yang et al. 2011).

The oxygen tolerance, denitrification rate, and denitrification products of some aerobic denitrifying bacteria have been studied in detail. As listed in Table 1.1, the majority of aerobic denitrifying bacteria has a good tolerance to oxygen and can maintain a high denitrification rate under aerobic conditions. However, since N_2O

Table 1.1 Aerobic denitrification characteristics and gaseous products of some aerobic denitrifying bacterial strains

Aerobic denitrifying bacteria strain	Culture medium	Oxygen concentration[a]	Nitrogen removal rate (mg L^{-1} h^{-1})	NO emission [b](%)	N_2O emission[b] (%)	References
Alcaligenes faecalis DSM30030	Ammonia-citrate	120 rpm	n.a.	0.31	26.17	Papen et al. (1989)
Paracoccus denitrificans LMD 92.63	Ammonia-acetate	<1–50%	n.a.	n.a.	76.47–92.71	Arts et al. (1995)
Paracoccus denitrificans LMD 92.63	Ammonia-nitrate-acetate	20–95%	n.a.	n.a.	95.79–100	Arts et al. (1995)
Alcaligenes faecalis TUD	Ammonia-nitrite-acetate	0–5%	0.138–0.92	n.a.	26–90	Otte et al. (1996)
Alcaligenes eutrophus	Ammonia-nitrate-acetate	0–1%	15.19 (NO3-N)	0.19	15	Kester et al. (1997)
Pseudomonas stutzeri	Ammonia-nitrate-acetate	0–1%	7.56 (nitrate nitrogen)	0.13	0.47	Kester et al. (1997)
Unknown strain 1	Nitrite-acetate-yeast extract	7.2–6.0 mg L^{-1}	n.a.	n.a.	80	Frette et al. (1997)
Pseudomonas putida WXZ-4	Nitrite-succinate	66–91%	0.69 (nitrate nitrogen)	n.a.	n.a.	Wang et al. (2007)
Delftia tsuruhatensis WXZ-9	Nitrate-succinate	66–91%	0.73 (nitrate nitrogen)	n.a.	n.a.	Wang et al. (2007)
Agrobacterium tumefaciens C58	Ammonia-nitrite-succinate	7–0%	n.a.	78.26	21.73	Bergaust et al. (2008)
Agrobacterium tumefaciens C58	Ammonia-nitrate-succinate	7–0%	n.a.	16.67	83.33	Bergaust et al. (2008)
Paracoccus denitrificans ATCC 35512	Livestock wastewater	0–3%	n.a.	n.a.	80	Miyahara et al. (2010)
Pseudomonas stutzeri TR2	Livestock wastewater	0–3%	n.a.	n.a.	n.d.	Miyahara et al. (2010)

Pseudomonas sp. yy7	Nitrite-acetate	0–7.5 mg L^{-1}	0.76 (nitrite)	n.a.	n.a.	Wan et al. (2011)
Bacillus methylotrophicus L7	Nitrite-succinate	160 rpm	0.24 (nitrite)	n.a.	n.q.	Zhang et al. (2012)
Rhodococcus sp. CPZ24	Nitrate-succinate	120 rpm	0.93 (nitrate)	n.a.	n.a.	Chen et al. (2012)
Acinetobacter sp. HA2	Nitrite-acetate or nitrate-acetate	150 rpm	2.51 (nitrite) 1.88 (nitrate)	n.a.	n.a.	Yao et al. (2013)
Aeromonas sp. HN-02	Nitrate-succinate	160 rpm	3.55 (nitrate)	n.a.	n.a.	Chen et al. (2014)
Chryseobacterium sp. R31	Nitrate-glucose	100 rpm	1.0 (nitrate)	n.a.	n.a.	Kundu et al. (2014)
Pseudomonas stutzeri T13	Nitrate-succinate or nitrite-succinate	160 rpm	12.40 (nitrate) 6.53 (nitrite)	n.a.	n.a.	Sun et al. (2015)
Vibrio diabolicus SF16.	Nitrate-acetate	120 rpm	8.53 (nitrate)	n.a.	n.a.	Duan et al. (2015)

[a]rpm represents rotations per minute, % represents degree of saturation
[b]percentage of NO or N^2O accounting for the removed nitrogen
n.a. *not available*, n.q. *not quanlified*

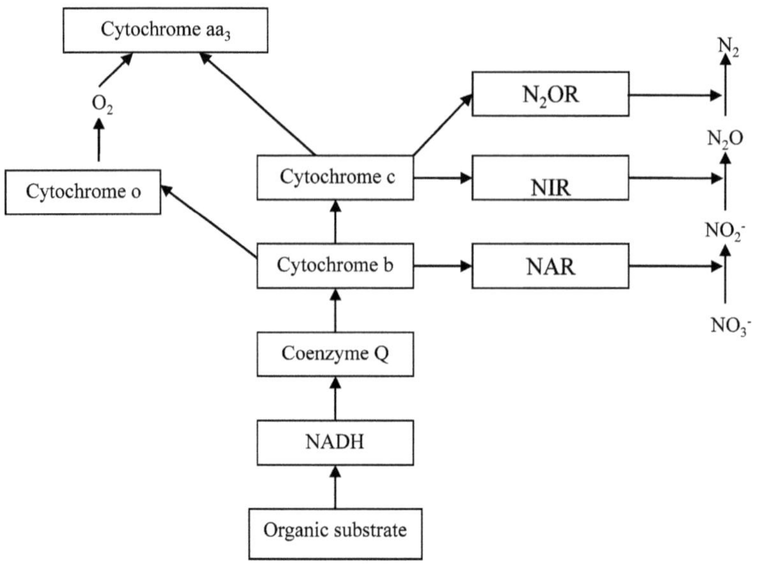

Fig. 1.2 Electron transfer model in the aerobic denitrifying process

reductase is very sensitive to O_2 and can be completely deactivated even in the presence of trace amounts of O_2, this led to the generation of large amounts of N_2O emissions by most aerobic denitrifying bacteria (Chen and Strous 2013). In addition, some heterotrophic bacteria do not have a complete denitrifying enzyme system, such as *Agrobacterium tumefaciens* lack N_2O reductase gene, resulting in N_2O instead of N_2 as the denitrification end product (Bergaust et al. 2008).

To explain the aerobic denitrification phenomenon, Robertson et al. (1988) proposed a "synergistic breath" theory that the excess reducing power is reduced by the periplasmic nitrate reductase, converting nitrate to nitrate, as the result of the "bottleneck" of electron transfer to oxygen atom, which explains the oxygen denitrification phenomenon that oxygen and nitrate can act as electron acceptors at the same time. Studies have shown that the lack of nitrate or O_2 will slow down the growth of bacteria and denitrification rate, and synergistic respiration can overcome the cytochrome c and cytochrome aa3 between the electron transport chain bottlenecks, allowing electronic flow simultaneously transmitted to the anti-nitrification enzymes and oxygen. Lloyd (1993) presents an electron transport model for the aerobic denitrification of bacteria, as shown in Fig. 1.2. Nitrate and O_2 can be used as the final electron acceptor. Denitrifying bacteria can transport electrons from the material being reduced to O_2 but also through the nitrate reductase electron transfer to nitrate nitrogen. In addition, many in-depth interpretation of the aerobic denitrification phenomenon were carried out through the related key enzymes studies.

1.2.2 Key Enzymes and Genes of Aerobic Denitrifying Bacteria

The traditional denitrification theory suggests that the existence of oxygen can inhibit the denitrification enzyme synthesis or denitrifying enzyme activity, and in the process of denitrification, oxygen will compete with nitrate for electrons, so denitrification can only be carried out in anoxic environment. In recent years, it has been found that some denitrifying enzymes can be synthesized and function under aerobic conditions, and even some denitrifying enzyme activities of bacteria are not affected by dissolved oxygen, so that oxygen and nitric nitrogen can be simultaneously used as electron acceptor. The key enzymes involved in aerobic denitrification mainly include nitrate reductase (NAR), nitrite reductase (NIR), nitric oxide reductase (NOR), and nitrous oxide reductase (N_2OR).

1. Nitrate Reductase (NAR)

Nitrate reductase catalyzes the first step of the denitrification process, the process of nitrate reduction to nitrite. There are two different types of nitrate reductase, i.e., membranous nitrate reductase (NAR) and periplasmic nitrate reductase (NAP). Studies have shown that the periplasmic nitrate reductase consists of a 90 kDa molybdopterin-binding catalytic subunit (NapA) and a 16 kDa cytochrome c552 (NapB), which contains two heme groups that catalyze the nitrates. The reduced active site is located on the NapA subunit (Bell et al. 1990). The genes encoding periplasmic nitrate reductase include the napEDABC gene cluster, which is responsible for protein synthesis or electron transport (Sears et al. 2000). Bell et al. (1990) cultured aerobic denitrifying bacteria *Thiosphaera pantotropha* under aerobic and anaerobic conditions and studied nitrate reductase and periplasmic nitrate reductase of bacteria under two conditions. Membrane nitrate reductase was active only in anaerobic environment and had denitrification before periplasm nitrate reductase. In aerobic environment, membrane nitrate reductase was restricted, while nitrate reductase showed activity and denitrification. Aerobic denitrification was associated with the expression of nitrate reductase in the periplasm, instead of the nitrate reductase normally expressed by anaerobic denitrifying bacteria. Oxygen is thought to be involved in regulating nitrate respiration by preventing nitrate from entering the catalytic center of membrane nitrate reductase, but the periplasmic nitrate reductase has a catalytically active center outside the cell membrane and therefore oxygen transport to nitrates and nitrites.

2. Nitrite Reductase (NIR)

Nitrite reductase is the key enzyme of the denitrification pathway. The reduction reaction of nitrite to NO is a marked reaction between denitrification and other nitrate metabolism. The expressed gene of this enzyme has been used as a molecular marker for denitrifying bacteria for their population structure and diversity (Braker et al.

2000). Nitrite reductase distributed in the outer membrane of the cell membrane including two types: cytochrome cd1-type nitrite reductase containing a c-type and d1-type heme encoded by the *nirS* gene and Cu-type nitrite reductase encoded by the *nirK* gene. Moir et al. (1993) purified the cytochrome cd1-type nitrite reductase from an aerobic denitrifying bacterium *P. pantotropha* with a molecular weight of about 63 kDa. Fulop et al. (1995) further found that the enzyme has two different catalytic functions, which can both catalyze the nitrite to get an electron to NO and catalyze O_2 to get 4 electrons to water.

3. Nitric Oxide Reductase (NOR)

NO reductase plays a key role in the process of denitrification. Fujiwara and Fukumori (1996) isolate NOR from aerobic denitrifying bacterial strain *P. pantotropha* ATCC 35512 and demonstrated that it participates in the aerobic denitrification process. The enzyme is located on the cell membrane, composed of two subunits with the molecular weight of 34 kDa and 15 kDa, respectively; each molecule contains two heme b and a heme c and norCB is the coding gene; without copper atom, a nonheme iron atom contains 1–2 g mol^{-1}. A high oxygen reductase activity manifests under the condition that the ferrous cytochrome c550 is an electron donor, and thus it is presumed that it is stable under aerobic conditions. NO reductase has a high affinity for NO; it can be limited to the concentration of electrons for NO reduction, thereby reducing the concentration of NO maintained at a very low level and to avoid the body itself caused by toxic effects.

4. N_2O Reductase (N_2OR)

The final step of the denitrification process N_2O reductase catalysis is the reduction of N_2O to N_2. Berks et al. (1993) purified N_2O reductase from aerobic and anaerobic T. pantotropha cells, respectively, and found that the molecular characteristics of N_2O reductase were similar to N_2O reductase extracted from anaerobic denitrifying microorganisms. The enzyme can be 1% acetylene and 0.125 mmol L^{-1} cyanide completely inhibited. N_2OR is located in the periplasm of the cell membrane and is a copper dimer with a molecular weight of 67 kDa. It has two poly-Cu centers, Cu_A and Cu_Z. Cu_A get electrons from the cytochrome c or pseudocryptin and transfer to Cu_Z. Since Cu_Z can be deactivated by trace amounts of O_2, the N_2O reduction process of most aerobic denitrifiers is inhibited, resulting in a large amount of N_2O emission (Arts et al. 1995). N_2OR is encoded by the nos gene cluster. The denitrifying bacteria have little change in the gene cluster, which is usually composed of nosL, nosY, nosF, nosD, nosZ, and nosR (Philippot et al. 2001).

1.2.3 Application of Aerobic Denitrifying Bacteria in Wastewater Treatment

Since the 1990s, the application of aerobic denitrifying bacteria in the process of enhanced biological nitrogen removal has attracted extensive attentions. It is an effective way to combine the aerobic denitrifying bacteria with nitrifying bacteria or activated sludge to strengthen aerobic nitrogen removal capability of the wastewater treatment system. It was found that the addition of aerobic denitrifying bacteria could achieve simultaneous nitrification and denitrification in the biological denitrification system with the advantages of no requirement of neutralization, strong anti-shock capability, and high denitrification efficiency. The outstanding excellence compared with the traditional wastewater treatment methods laid the foundation for the future engineering applications.

Kshirsagar et al. (1995) applied the aerobic denitrification bacterium *T. pantotropha* to the activated sludge system to treat nitrogen-containing wastewater and found that the total nitrogen removal rate was 20% higher than that of the unenhanced system. Gupta and Gupta (2001) carried out the pure culture of aerobic denitrifying bacteria *T. pantotropha*, which was inoculated into a rotary biofilm reactor to treat nitrogen-containing wastewater. After the bacteria was attached to the membrane, the reaction system worked well and could simultaneously remove high concentrations of ammonia nitrogen and total nitrogen and no significant accumulation of nitrite and nitrate. Patureau et al. (2000) fixed the aerobic denitrifying bacterium *M. aerodenitrificans* in alginate microspheres and mixed with activated sludge system in SBR reactor. By controlling dissolved oxygen and adding carbon source, the stain survived and ultimately performed nitrogen and phosphorus removal capability in the complex environments. Joo et al. (2006) used heterotrophic nitrification-aerobic denitrification bacteria *A. faecalis* no.4 to treat aquaculture wastewater. After continuous aerobic incubation, the COD of 2000 mg L^{-1} and 12,000 mg L^{-1} in wastewater was almost completely removed, and 65% of the ammonia nitrogen was removed by bacterial heterotrophic nitrification and aerobic denitrification. Yang et al. (2011) screened the aerobic denitrifying bacteria *B. subtilis* A1 into the SBR reactor to strengthen the treatment of nitrogen-containing wastewater and found that aerobic denitrifying bacteria after the SBR aerobic section generated nitric nitrogen was significantly reduced. The total nitrogen removal rate can be increased from 61.2% to 81.3%. Chen et al. (2015) inoculated aerobic denitrifier *Pseudomonas* sp. X31 into a bio-ceramic reactor to obtain more than 95% nitrification efficiency. Duan et al. (2015) showed that the newly selected aerobic denitrifying bacterium *Vibrio diabolicus* SF16 was inoculated into the membrane bioreactor to treat salt-containing wastewater. The removal rates of ammonia nitrogen and total nitrogen reached 97.14% and 73.92%, respectively, and showed good engineering application potential.

1.3 Nitrogen Oxides in the Biological Nitrogen Removal (BNR) Process

1.3.1 Nitrogen Oxides in the Biological Nitrogen Removal Process

Nitrogen oxides include a variety of compounds such as nitrous oxide (N_2O), nitric oxide (NO), nitrogen dioxide (NO_2), dinitrogen trioxide (N_2O_3), dinitrogen tetroxide (N_2O_4), and dinitrogen pentoxide (N_2O_5). N_2O is one of the three major greenhouse gases in the atmosphere after CO_2 and CH_4, and its greenhouse effect is 300 times over that of CO_2. The N_2O concentration in the atmosphere has increased from 270 ppb in 1750 to 319 ppb in 2005, and the contribution of N_2O to global warming has reached 7.9% in 2004 (IPCC 2007). Some studies have suggested that N_2O will be the most important substance destroying the ozone layer in the twenty-first century (Ravishankara et al. 2009). The Intergovernmental Panel on Climate Change (IPCC), co-established by the United Nations Environment Program (UNEP) and the World Meteorological Organization (WMO), has identified N_2O as a major problem affecting natural ecosystems and threatening the foundations of human existence. In addition, the NO and NO_2 (NOx) in the air are one of the important pollutants in the atmosphere, which can cause serious environmental problems such as acid rain, photochemical smog, and ozone hole (Barnes et al. 1995). However, since the biological denitrification process is not the main source, therefore, we will focus on N_2O emissions.

Human activities contribute 40–50% of annual N_2O enrichment, 80% of which comes from agricultural production. Other sources include biomass combustion, organic and nitric acid production, and solid waste treatment (Dalal et al. 2003; IPCC 2007). Recent studies have shown that sewage treatment systems are an important source of N_2O emissions, and emissions are increasing. For the first time, Czepiel et al. (1995) reported a detailed study of N_2O emissions from the Durham WWTP in New Hampshire, USA, and found that the total annual N_2O emissions from the wastewater treatment plant was 1.1×10^{10} ton, with a per capita annual emission factor of about 3.2 g. Hereafter N_2O emissions from the sewage treatment facilities have caused a wide range of people's attention.

Because of the complexity of measuring N_2O emissions in on-site wastewater treatment facilities, there is no standardized method for the measurement of N_2O emissions. The total amount of N_2O emissions from wastewater treatment plants is therefore estimated based primarily on the model. According to the US EPA and IPCC study (IPCC 2007), N_2O emissions from wastewater treatment facilities account for 3% of all human activities and sixth of all sources. It is estimated that by 2020, the global sewage treatment plant N_2O emissions will grow 13% compared to 2005.

The N_2O emission from wastewater treatment plants mainly occurs during the biological denitrification (BNR) process, which is an intermediate product of denitrification process and a by-product of ammonia-oxidizing bacteria (AOB). The biological denitrification mechanism of wastewater is the same as that of biological nitrogen removal in soil, river, and ocean. However, the sewage treatment process has a high rate of nitrogen conversion, and the biological denitrification characteristic in other environment differs. First, the total nitrogen concentration of the effluent in the general sewage plant is 20–70 mg L^{-1}, and the nitrogen conversion can be completed in 3–8 h, so the nitrification and denitrification rate is high. The N_2O emission rate is also closely related; second, in the biological sewage treatment process, the microorganisms will undergo a short period of aerobic, anaerobic environment conversion, so rapid environmental changes on the nitrification/denitrification of the physiological metabolism of bacteria are a huge challenge; again, the aeration system is an important equipment in the biological sewage treatment system, the role of the nitrification process is huge, but also N_2O from liquid phase into the gas phase is the most important factor; finally, the sewage treatment process is built by the artificial operation of the system that can be optimized by optimizing the process design and operating conditions to reduce N_2O emissions.

1.3.2 *Determination of Nitrogen Oxides Concentrations*

1. Determination of N2O Concentration in Gas Phase

In actual wastewater treatment plants, the N_2O gas emitted from the sludge system is typically collected by means of a closed device floating on the surface, which was originally used to collect the gases from the solid surface. Czepiel et al. (1995) used this method when studying N_2O emissions from the Durham Sewage Treatment Plant in New Hampshire, USA. Aerobic zone due to the presence of aeration device, dissolved N_2O in the liquid phase, can be blown out into the collection device and in the anoxic area mainly through the exchange of gas-liquid two-phase collection. Because of the lack of on-site early measurement methods, the use of closed-device collection of N_2O gas is generally through the airbag or syringe collection and storage and then to the laboratory for analysis.

Although the N_2O gas released from the liquid phase can be successfully collected by means of a closed device, the intermittent sampling is not easy to obtain the dynamic change of the N_2O emission rate, and the measurement result is inaccurate. In recent years, continuous online measurement has been widely used. Among them, the online detectors include electron capture detectors (Otte et al. 1996), infrared analyzers (Ahn et al. 2010; Butler et al. 2009), Fourier transform infrared spectroscopy (Joss et al. 2009), chemiluminescence analyzer (Kampschreur et al. 2008a), and mass spectrometry (Zeng et al). In these methods, the electron capture detector and infrared analyzer can measure up to 2000 ppm concentration of N_2O gas; the most widely used, chemical fluorescence analyzer, the highest sensitivity, can be used to analyze trace concentrations of N_2O gas.

In addition to studying the rate of change of emissions over time, spatial variations also have a significant effect on N_2O gas emissions in some wastewater treatment processes, such as plug flow reactors and oxidation ditch reactors (Aboobakar et al. 2013). It is important to simultaneously detect the emission characteristics of N_2O gas in several regions.

2. Measurement of N_2O Concentration in Liquid

In the early laboratory study, N_2O concentration in the liquid phase was measured by field sampling and then returned to the laboratory for analysis. In the field, a liquid sample with a volume less than the container volume was taken with a syringe or headspace bottle. The N_2O concentration in the gas phase was measured after the gas-liquid phase reached equilibrium.

In recent years, the use of microelectrodes for continuous measurement of N_2O concentration in liquid phase has been widely used, which is the most widely used Clark microelectrode (Unisense, Denmark) (Kampscherur 2008b; Foley et al.). The electrode in the 0–1.2 mM between the output signals can be maintained with a good linear relationship, and along with its low detection limit, its range of applications and reliability is greatly increased.

The N_2O concentration in the liquid phase is also closely related to the sampling position, so it is important to study the spatial distribution of N_2O concentration in the biological treatment process. Other operating parameters such as pH, dissolved oxygen, temperature, and sludge concentration are measured together for the purpose of mass balance calculation, model analysis, etc., while measuring N_2O concentration.

1.3.3 Emission Factors of Nitrogen Oxides in BNR Processes

N_2O emissions during wastewater treatment are usually expressed as the ratio of the total amount of N_2O emitted to the total nitrogen or total nitrogen removed in the biological treatment process, which is called the N_2O emission factor. It is clear from the study of N_2O in wastewater treatment (Tables 1.2 and 1.3) that the emission factor values vary widely, with emission factors ranging from 0% to 25% in the on-site wastewater treatment process. In the laboratory study, it is as high as 0–90%. The reason for this phenomenon is the first treatment process and operating conditions are different, followed by sampling and quantitative methods of difference. But such a high N_2O range also shows that controlling the process conditions and

Table 1.2 N_2O emission from biological treatment system of real wastewater treatment plants

Process type	N_2O emission factor	Sampling methodology	Main conclusion	References
AS (11,000 p.e.)	0.035%	Instantaneous sampling (once a week for 15 weeks)	N_2O was mainly produced in the aeration. The annual per capita N_2O emissions were3.2 g a^{-1} p^{-1} in urban sewage treatment plants	Czepiel et al. (1995)
AS (60,000 p.e.)	0.001%	Instantaneous sampling (once every 2 weeks for 1 year)	The N_2O emission increases with the increase of the concentration of nitrite and nitrate in the influent	Sumer et al. (1995)
AS (1000 p. e.)	0.01–0.08%	Online monitoring (4 aeration cycles for 2 h)	N_2O emissions are reduced as the aeration time decreases	Kimochi et al. (1998)
Partial nitrification-anammox	2.30%	Online monitoring (For 4 days)	N_2O emission in the aerobic stage increases with the decrease of dissolved oxygen. In the hypoxic stage N_2O increases with the increase of nitrite	Kampschreur et al. (2008b)
SBR partial nitrification-anammox n	0.4–0.6%	Online monitoring	N_2O emissions are slightly higher than the normal denitrification process	Joss et al. (2009)
BNR (12 sewage plants)	0.003–2.59%	Online monitoring	The contribution of aerobic segments to total N_2O emissions is significantly higher than that of hypoxia	Ahn et al. (2010)
BNR (7 sewage plants)	0.6–25.3% (均值3.5%)	Instantaneous sampling	N_2O emissions and process type and nitrite concentration are closely related	Foley et al. (2010)
Partial nitrification-anammox	5.1–6.6%	Online monitoring	The higher N_2O emissions are partly due to the individual short-range nitrification stages	Desloover et al. (2011)

Engineering type	N_2O release factor	Selection methodology	Conclusion references	References
Push flow reactor (210,000 p.e.)	Gaseous average 0.036% Liquid average 0.01%	Online monitoring	N_2O emissions were negatively correlated with DO concentration	Aboobakar et al. (2013)
Aerobic – anoxic reactor	0.12–0.41%	Online monitoring	N_2O emissions decreased with increasing DO concentration and SRT, and the impact load capacity increased	Lotito et al. (2012)
Improved UCT process (750,000 p.e.)	0.96%	Online monitoring	Nitrification contributes significantly to N_2O; the rainy season is greater than that in the dry season	Guo et al. (2013)

(continued)

Table 1.2 (continued)

Engineering type	N_2O release factor	Selection methodology	Conclusion references	References
Nitrification SBR process (80,000 Ton/year)	5.60%	Instantaneous sampling	The low dissolved oxygen in the nitrification process is the main factor leading to N_2O emissions	Sun et al. (2013)
Oxidation ditch process (6000 Ton/day)	$0.52 \pm 0.16\%$	Online monitoring	The surface aeration area is the main area of N_2O emissions	Ye et al. (2014)

Note: *p.e.* population equivalent, *AS* activated sludge engineering, *BNR* desulfurization engineering, *SBR* introduction activated sludge engineering

optimization can be effective in reducing N_2O emissions. Foley et al. (2010) concluded that a wastewater treatment plant with a higher denitrification efficiency than a nitrifier-only wastewater treatment plant has a lower N_2O emission, demonstrating the feasibility of simultaneously improving the nitrogen removal efficiency of the wastewater treatment process and reducing N_2O emissions.

It can be seen from Tables 1.2 and 1.3 that N_2O can be produced in the process of nitrification or denitrification in biological wastewater treatment process. Therefore in the sewage treatment process, researchers generally only pay attention to these two biochemical reaction zones. Czepiel et al. (1995) monitored N_2O emissions at different stages of the wastewater treatment process. Ninety percent of the emissions were generated in the biochemical reaction stage, 5% in the grit chamber, and 5% in the sludge storage tank. Ahn et al. (2010) found that emissions of N_2O in the aeration area are typically two to three orders of magnitude higher than in the non-aeration zone. In the biochemical stage, it is not difficult to understand that N_2O emissions occur mainly in the aeration zone in sewage treatment processes, as N_2O is mainly discharged into the atmosphere in gaseous form (Sommer et al. 1998). This does not mean that N_2O is mainly produced in the aeration zone, since the Henry's coefficient of N_2O in water is 24 mM atm^{-1} (25 °C) and the Henry's coefficient of O_2 at the same conditions is only 1.3 mM atm^{-1} (Dean 1992), indicating that N_2O is more easily dissolved in water, so its presence in the absence of aeration conditions can be accumulated in the water to a high concentration. Law et al. (2011) found that in the nitrification reactor, N_2O concentrations in the non-aeration zone can accumulate to 0.5 mg L^{-1} with almost no N_2O emissions, but in the aeration zone, N_2O dissolved in air is 0.01–0.03 mg L^{-1}.

In addition to gaseous N_2O emissions, dissolved N_2O, which is accompanied by effluent loss, eventually enters the receiving water body and leads to increased N_2O emissions from the receiving waters, but this partial emission factor has not received much attention from researchers. Sommer et al. (1998) suggested that the dissolved N_2O concentration in the winter wastewater treatment system is five times higher than that of the N_2O produced by the direct blowoff of N_2O, due to the low water temperature in winter and the concentration of dissolved N_2O. Aboobakar et al.

Table 1.3 N_2O emission from biological treatment system of lab-scale wastewater treatment plants

Process type	N_2O emission factor	Sample methodology	Main conclusions	References
AS (continuous + order batch) – artificial water distribution	0–8%	Instantaneous sampling	The N_2O emission increased with the decrease of COD/N value and the increase of nitrite concentration	Hanaki et al. (1992)
Anoxic-aerobic SBR (Continuous) – high load wastewater	1–35%	Instantaneous sampling	The longer the aeration time, the higher the N_2O emission, which is related to the increase of nitrite concentration	Osada et al. (1995)
DAS (batch) – actual wastewater	1.20%	Instantaneous sampling	N_2O emissions increased with decreasing pH	Thorn and Sorensson (1996)
AS (continuous, 50d) – actual wastewater	0.2–1.5%	Instantaneous sampling (once a week)	The N_2O emission increased with the decrease of dissolved oxygen concentration and SRT	Noda et al. (2003)
Anaerobic-anoxic SBR (continuous) – artificial water distribution	90%	Online monitoring	N_2O emissions increase with the increase of nitrite concentration; N_2O may be the main product of denitrifying bacteria	Zeng et al. (2003)
NAS (continuous) – artificial water distribution	0.2–0.5%	Instantaneous sampling (once a day)	The N_2O emission ratio increases with increasing salt concentration	Tsuneda et al. (2005)
NAS (batch, 5 h) – actual wastewater	0.1–0.4%	Online monitoring	The N_2O emission was the highest when the dissolved oxygen concentration was 1.0 mgO_2L^{-1}	Tallec et al. (2006)
Nitration SBR (continuous) – artificial water distribution	3%	Online monitoring	N_2O emissions increased with increasing nitrite concentration	Kampschreur et al. (2008b)
MABRs (continuous) – artificial water distribution	Aeration section~0.015%	Online monitoring	Intermittent aeration has successfully enriched	Pellicer-Nàcher et al. (2010)

(continued)

Table 1.3 (continued)

Process type	N_2O emission factor	Sample methodology	Main conclusions	References
	non-aeration section <0.001%		short-range nitrification-anaerobic ammonium oxide and greatly reduced N_2O emissions	
Anaerobic-microaerobic SBR	10.4–28.2%	Online monitoring	Sludge hydrolysis acidification solution can significantly reduce N_2O emissions	Zhu and Chen (2011)
AS (batch, 4 h) – artificial water distribution	0.2–18.9%	Online monitoring	N_2O emissions are related to nitrite nitrogen concentration; N_2O emission in denitrification process increases significantly with increasing dissolved oxygen	Wunderlin et al. (2012)
Aerobic granular sludge (continuous) – artificial water distribution	2.2–8.2%	Online monitoring	The N_2O emission decreases with the increase of the aeration rate and decreases with the increase of C/N ratio	Quan et al. (2012)
Anaerobic-microaerobic SBR	7.70%	Instantaneous sampling	Autotrophic denitrification is the main mechanism for N_2O production	Jia et al. (2013)
Nitrated granular sludge	5.41%	Instantaneous sampling	The nitrification process is the main source of N_2O production	Liang et al. (2014)
Nitration SBR	2.20%	Continuous sampling	N_2O was negatively correlated with DO in the range of 1–4.5 mgL^{-1} dissolved oxygen	Pijuan et al. (2014)

Note: *AS* activated sludge process, *NAS* nitrification activated sludge, *DAS* activated sludge

(2013) found that the gaseous N_2O emission factor was 0.036% of the total influent TN, and the dissolved N_2O emission factor with the effluent was 0.01%, based on a long-term continuous observation of an actual WWTP. Therefore, in addition to gaseous N_2O emissions, dissolved N_2O with effluent should also be of major interest.

1.4 Nitrogen Oxides Production and Control in BNR Process

1.4.1 N_2O Production Mechanism in the BNR Process

During the biological denitrification process, N2O is mainly produced as by-products or intermediates in autotrophic nitrification, autotrophic denitrification, and heterotrophic denitrification, as shown in Fig. 1.3.

1. Autotrophic Nitrification

Nitrification is the process by which the ammonia-oxidizing bacteria and nitrite-oxidizing bacteria transform the ammonia-nitrogen into nitrite-nitrogen in the wastewater. Ammonia-oxidizing bacteria first oxidize ammonia to hydroxylamine under the action of ammonia monooxygenase (AMO), then oxidize to nitrite by the action of hydroxylamine oxidoreductase (HAO), and obtain energy in two steps of oxidation process. Nitrite-oxidizing bacteria convert nitrite to nitrite and obtain energy under the action of nitrite oxidoreductase (NOR), both of which use inorganic carbon as the carbon source for their own growth and metabolism.

In nitrification, N_2O is mainly produced in the process of hydroxylamine hydroxyl oxidation. NOH or $N_2O_2H_2$ intermediates are formed by hydroxylamine oxidation, and because of its instability, the intermediates will be further hydrolyzed into N_2O (Sutka et al. 2006; Poughon et al. 2001). The experimental results of nitrifying sludge showed that the rate of N_2O production had a high correlation with the rate of ammonia oxidation. Law et al. (2012) found a good exponential relationship between the two, and the N_2O emissions were modeled by the chemical decomposition of NOH. The N_2O emission during the ammoxidation process was confirmed by simulation of the exponential relationship which supports the assumption of chemical decomposition of NOH. In addition to the chemical decomposition of NOH intermediates, hydroxylamines are chemically reacted with nitrites or O_2 to produce N_2O gases due to their strong reducing properties (Van Cleemput 1998), as shown in formulas (1.1) and (1.2); however, some researchers believe that during the normal biological treatment of wastewater, the release of N_2O is negligible due to the low content of hydroxylamine as an intermediate in the process of ammonia oxidation (Wunderlin et al. 2012).

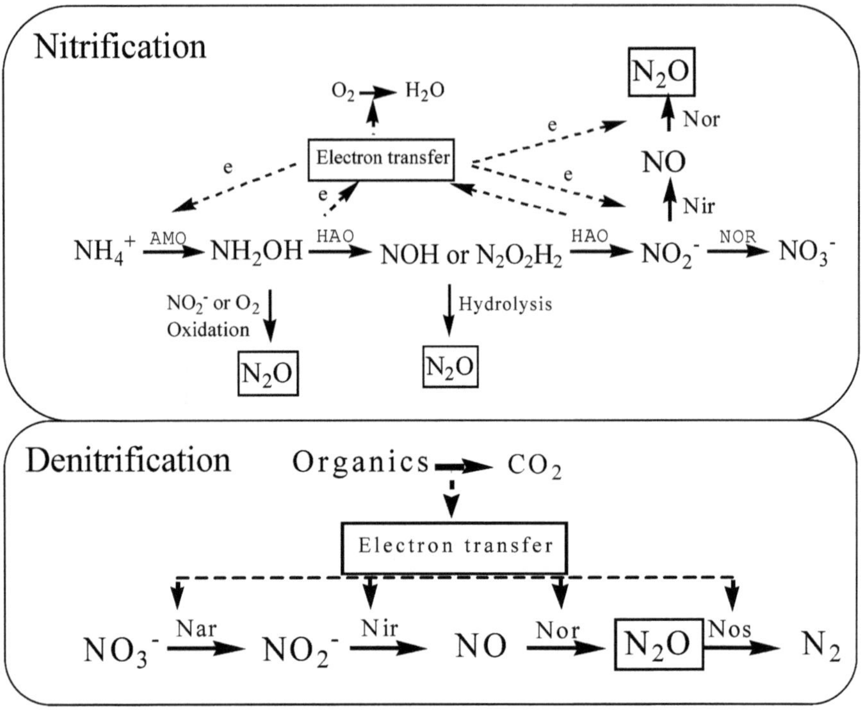

Fig. 1.3 N_2O generation mechanisms during biological nitrogen removal process

$$NH_2OH + 0.5\,O_2 \rightarrow 0.5\,N_2O + 1.5\,H_2O \tag{1.1}$$

$$NH_2OH + NO_2^- + H^+ \rightarrow N_2O + 2\,H_2O \tag{1.2}$$

Nitrification can also be done by some heterotrophic microorganisms, known as heterotrophic nitrification. Heterotrophic bacteria take organic carbon as growth and metabolism of energy and carbon source. Heterotrophic nitrification rates are 100–1000 times lower than autotrophic nitrification rates and are predominant only at high C/N ratios (COD/N > 10) and low dissolved oxygen concentrations (Van Niel et al. 1993). Although heterotrophic nitrification has not been reported as predominant in wastewater treatment, studies have shown that heterotrophic nitrifying bacteria produce more N_2O than autotrophic nitrifying bacteria (Anderson et al. 1993).

Recently, ammonium-oxidizing archaea (AOA), an ancient bacterium capable of ammoxidation, has been isolated (Konneke et al. 2005). Santoro et al. (2011) found that N_2O in the oceans is mainly produced by AOA rather than AOB. Ammonia-oxidizing archaebacteria have also been found in wastewater treatment processes operated at low dissolved oxygen and high sludge age (Park et al. 2006), but the process of ammonia oxidation of archaebacteria on N_2O emissions from wastewater denitrification process contribution has not been reported.

2. Autotrophic Denitrification

Autotrophic denitrification refers to the process by which ammonia-oxidizing bacteria reduce the nitrite to NO, N_2O, or N_2 with electron donor such as hydroxylamine, ammonia nitrogen, or hydrogen (Bock et al. 1995; Poth and Focht 1985). However, to date, only genes (nirK and norS) expressing nitrite reductase and NO reductase have been found in ammonia-oxidizing bacteria, and no N_2O reductase-expressing gene has been found. Therefore, when ammonia-oxidizing bacteria undergo denitrification, the final product is N_2O rather than N_2 (Beaumont et al. 2004; Cantera and Stein 2007; Casciotti and Ward 2001; Garbeva et al. 2007).

Autotrophic denitrification plays an important role in N_2O emission, especially when the surrounding environment is changed from hypoxia to aerobic state. Tallec et al. (2006) studied the mechanism of N2O production by controlling the nitrification and denitrification processes at different stages using different inhibitors, indicating that autotrophic denitrification accounts for 58–83% of total N_2O emissions in actual wastewater treatment processes. This value is affected by different dissolved oxygen. The results of Kim et al. (2010) also show that denitrification activity of ammonia-oxidizing bacteria plays a decisive role in N_2O emission in wastewater treatment, and this result is proved by detecting the expression of nitrite reductase (nirK).

3. Heterotrophic Denitrification

Denitrification is a process in which heterotrophic microorganisms use organic carbon as electron donor to reduce nitric nitrogen to nitrite, nitric oxide, nitrous oxide, and nitrogen under the condition of anaerobic or low dissolved oxygen. There are four main enzymes in the process of denitrification, which are nitrate reductase, nitrite reductase, nitric oxide reductase, and nitrous oxide reductase. Since N_2O is an intermediate product of the denitrification process, any N_2O reductase-inhibiting factor is responsible for the accumulation of N_2O. In recent years, aerobic denitrifying bacteria have received extensive attention, as it can reduce nitrate nitrogen to gaseous products under anaerobic or aerobic conditions. However, due to the susceptibility to dissolved oxygen of nitrous oxide reductase activity, the majority of final gaseous product of bacteria is dominated by N_2O (Korner and Zumft 1989).

1.4.2 *Influencing Factors of N_2O Emission from BNR Process*

1. Dissolved Oxygen

In the biological denitrification process, dissolved oxygen concentration has an important effect on both nitrogen conversion efficiency and N_2O emission during nitrification and denitrification (Kampschreur et al. 2008a, b; Zheng et al. 1994).

Most studies have shown that low dissolved oxygen conditions are responsible for the increase in N_2O emissions during nitrification (Tallec et al. 2006). Nitrification requires aerobic conditions, which generally require dissolved oxygen concentration higher than 1.0 mg L^{-1} so that nitrification reaction can be more complete, or it will seriously affect the rate of nitrification. The sensitivity of different nitrifying bacteria to the concentration of dissolved oxygen is different. Generally, the ammonia-oxidizing bacteria have a dissolved oxygen half-saturation coefficient of 0.2–0.4 mg L^{-1} and nitrite-oxidizing bacteria of 1.2–1.5 mg L^{-1}. Thus, when dissolved oxygen is insufficient, the nitrite-oxidizing bacteria are not dominant in the competition with the dissolved oxygen of the ammonia-oxidizing bacteria, resulting in the nitrite oxidation rate being lower than the rate of ammonia oxidation and the accumulation of nitrite nitrogen (Park et al. 2000). Ammonia-oxidizing bacteria also use nitrite instead of oxygen as an electron acceptor for the purpose of oxygen-saving ammonia oxidation process, thus accelerating the N_2O emission rate.

Hu et al. (2011) studied the effect of aeration on N_2O production by using SBR reactor. It was found that the N_2O emission decreased with the increase of aeration amount. When the aeration amount was 1 L h^{-1}, the amount of N_2O produced in the aerobic stage was 11.51 mg, and when the aeration increased to 4.4 L h^{-1}, the value decreased to 2.92 mg. Foley et al. (2010) investigated the emission of N_2O in an oxidation ditch process wastewater treatment project and found that N_2O production increased with the decrease of oxygenation efficiency after aeration brushing. Yu et al. (2010) further studied the growth and metabolism of ammonia-oxidizing bacteria under low dissolved oxygen and nitrite accumulation at the level of gene expression and transcription. It was found that the mRNA content of AMO and HAO in the bacteria was lower than that in the logarithmic growth phase under the lower dissolved oxygen concentration. The results showed that the mRNA content of AMO and HAO was lower than that of the control. Nitrite reductase (Nir) and NO reductase (Nor) in the culture medium contain 280 mg L^{-1} nitrite. Therefore, they concluded that when the dissolved oxygen content is limited, the growth and metabolism of *N. europaea* will be adjusted to increase the rate of ammonia and hydroxylamine oxidation. When the content of nitrite increases, the bacteria will reduce the synthesis of denitrifying enzymes nitrous acid accumulation in order to reduce the toxicity of their own growth, and both cases will lead to an increase in N_2O emissions. Ahn et al. (2010) investigated N_2O emissions from 12 wastewater treatment plants in the United States. It was found that N_2O emissions in the aerobic zone were also positively correlated with dissolved oxygen (DO) and dissolved oxygen (DO) levels in addition to the anaerobic zone. The researchers pointed out that the rapid increase in dissolved oxygen will lead to the accumulation of nitrite nitrogen which leads to an increase in N_2O emissions. Kampschreur et al. (2009) found that the amount of N_2O produced by ammonia-oxidizing bacteria decreased with the decrease of dissolved oxygen in the nitrification-anaerobic ammonium oxidation process, but the mechanism of this phenomenon needs further proof and explanation.

Denitrification generally requires less than 0.5 mg L^{-1} dissolved oxygen, while excessive dissolved oxygen will inhibit the denitrification enzyme synthesis and activity. Particularly N_2O reductase is more sensitive to O_2, resulting in denitrification process being incomplete and eventually leading to an increase in N_2O emissions. Studies have shown that very low concentrations of dissolved oxygen also have a significant inhibitory effect on N_2O reductase activity, even in novel aerobic denitrification processes (Otte et al. 1996). Lu and Kartik (2010) studied the changes of biological nitrogen removal and N_2O production with different dissolved oxygen concentrations using SBR reactor. Nitrate was used as nitrogen source and ethanol as carbon source. Three dissolved oxygen concentrations were set at 2, 5, and 9 mg L^{-1}. It was found that with the increase of dissolved oxygen concentration, the removal rate of total nitrogen decreased and the N_2O emission increased. At the dissolved oxygen concentration of 9 mg L^{-1}, the SBR system had the highest N_2O emission which is 7.1% of the total nitrogen load. Tallec et al. (2006) used a small-scale experiment to study the N_2O production at a dissolved oxygen concentration in a Paris wastewater treatment plant and found that N_2O emissions were the highest at a dissolved oxygen concentration of 1.0 mg L^{-1} and 7.1 μg N_2O in average. The contribution of nitrification and denitrification to N_2O emissions at a low dissolved oxygen concentration (0.1–2 mg L^{-1}) was obtained by the inhibition of nitrification with the inhibitor allylthiourea (ATU) and sodium chlorate, N g^{-1}SS h^{-1}. The N_2O emission from autotrophic denitrification was found to be 58–83% of total N_2O emission. As the dissolved oxygen concentration increased, the ratio of heterotrophic denitrification to N_2O emission decrease from 42% to 17%.

Biological wastewater treatment process generally sets up aerobic tank and anoxic tank. The same set of sludge system through the cycle between the reaction tanks completes the process of nitrification and denitrification, which shows that the sludge will continue in the alternating aerobic and anoxic environment, while rapid changes in dissolved oxygen can lead to a sharp increase in N_2O emissions (Kester et al. 1997; Yu et al. 2010). It was found that the maximum amount of N_2O produced by the autotrophic denitrification of ammonia-oxidizing bacteria was 0.1–0.3 mg L^{-1} for dissolved oxygen (Chuang et al. 2007; Goreau et al. 1980; Tallec et al. 2008; Wunderlin et al. 2012). Therefore, when the concentration of dissolved oxygen in the sludge environment changes, the N_2O produced by heterotrophic denitrification also increases with the concentration of dissolved oxygen in the heterotrophic denitrifying bacteria rising. N_2O reductase synthesis needs a longer delay time compared to nitrite reductase when there is sludge environment from aerobic to anoxic environment. In addition, when transformed from anoxic aerobic environment, N_2O reductase activity will be lost immediately, while nitrite reductase will remain a few hours of activity, though the activity and hypoxic conditions decreased. Kampschreur et al. (2008a) studied NO and N_2O emission rates of nitrified sludge when dissolved oxygen was changed. It was found that NO and N_2O emissions increased sharply when nitrogen was used to blow off the dissolved oxygen in the reactor and took seven or five times longer than the normal operation time. This phenomenon may be caused by the rapid growth of bacteria metabolism of the environment, which requires a certain amount of time to adapt to rapid changes in the activity of different enzymes leading to an imbalance between the increases in N_2O production.

2. Nitrite Concentration

Nitrite can not only cause toxic effects to the human body and aquatic plants but also affect the N_2O emission during biological treatment. The results show that nitrite accumulation can promote the N_2O emission in both autotrophic denitrification and heterotrophic denitrification processes. Excessive nitrite concentration first increases the rate of autotrophic denitrification, leading to an increase in N_2O emissions, followed by a reduction in N_2O reductase activity during heterotrophic denitrification, and further reduction of N_2O to enhance its emissions the amount.

Tallec et al. (2006) used an inhibitor to study the amount of N_2O produced in different biochemical processes under different dissolved oxygen conditions. When dissolved oxygen was 1.0 mg L^{-1} under the condition of 10 mg L^{-1} nitrite, the amount of N_2O produced by autotrophic bacteria increased eightfold, which was increased by fourfold when dissolved oxygen was 0.1 mg L^{-1}. Kampschreur (2008a) also demonstrated a significant increase in NO and N_2O production from the enriched ammonia-oxidizing bacteria each time when a certain amount of nitrite was added to the system. Foley et al. (2010) found that the N_2O production was positively correlated with the concentration of nitrite in the reactor through the analysis of the N_2O emission from the seven on-site wastewater treatment plants.

The effect of nitrite concentration on N_2O emission has been recognized by most researchers, but the mechanism of action has not been agreed. Schulthess et al. (1994) used continuous aeration and serum bottle experiments to monitor the gaseous intermediate NO and N_2O during denitrification. It was found that in the process of nitrification and denitrification, the accumulation of NO and N_2O has been inhibited. However, further studies have found that the inhibition of denitrifying enzymes is due to the accumulation of NO but not the accumulation of nitrite, and the artificial feeding of NO in serum bottles proved this inhibition. Zhou et al. (2008) studied the inhibitory effect of nitrite concentration on N_2O reduction at different pH values using denitrifying phosphorus removal sludge. It was found that the inhibition had a high exponential correlation with free nitrite concentration, and nitric acid, but not nitrite, is an inhibitor of N_2O reductase; at a concentration of free nitrite of 0.0007–0.001 mg L^{-1} (approximately equal to the nitrite concentration of 3–4 mg L^{-1} at pH 7.0), N_2O reductase activity was inhibited by about 50%, and when the nitrite concentration reached 0.004 mg L^{-1}, the activity was completely inhibited; further experiments found that N_2O reductase can be inhibited after a certain time to be restored, and recovery time and inhibition time, and only with the concentration of free nitric acid.

3. The Ratio of COD to Nitrogen

The COD/N has a great influence on N_2O production in denitrification process. It is generally believed that denitrification can occur only when the ratio of carbon to nitrogen is higher than 4. When the ratio of carbon to nitrogen is insufficient, the enzymes in the denitrification process such as nitrate reductase, nitrite reductase, NO reductase, and N_2O reductase will complete for electrons, which would increase N_2O production due to the weak competitiveness of N_2O reductase. Inadequate organic carbon sources are not the direct cause of N_2O emissions but rather due to the fact that denitrification does not completely lead to the accumulation of nitrite, which in turn leads to autotrophic and heterotrophic denitrification to produce N_2O (Hanaki et al. 1992). At present, the ratio of COD and N of domestic sewage in many large- and medium-sized cities in China is very low, so it is difficult to meet the demand of denitrification process for carbon source. Therefore, it is important to study the carbon-nitrogen ratio in wastewater biological treatment.

Hanaki et al. (1992) studied the amount of N_2O produced during denitrification at the C/N ratios of 1.5, 2.5, 3.5, and 4.5, respectively. It was found that the maximum N_2O emission was achieved at the C/N ratio of 1.5, reaching 10% of the influent nitrogen load. Itokawa et al. (2001) used intermittent aeration bioreactor to treat high pollutant load wastewater and used ^{15}N isotope tracer to study the N_2O emission during denitrification. It was found that when the ratio of carbon to nitrogen was lower than 3.5, N_2O load of 20–30% of the input water was discharged in the form of N_2O gas. Further research indicated that the N_2O gas emission mainly came from denitrification of nitrite during the second half of the hypoxic process. Schalk-Otte et al. (2000) studied the N_2O production of denitrifying bacteria *A. faecalis* at different carbon sources and found that N_2O production accounts for 1.8–10.4% of N load when the carbon source is sufficient. In the case of deficiency, the value is 32–64%. Schalk-Otte et al. (2000) used intermittent aeration bioreactor to treat domestic wastewater. It was found that when the influent carbon/nitrogen ratio (TOC/TN) was 2.6, 4.57% of the total influent nitrogen was discharged as N_2O gas. When methanol was added, and the influent C/N ratio was increased to 5.0, the value was only 0.14%. In biofilm reactor, the ratio of total nitrogen into N_2O gas was 3.01% and 0.12%, respectively. Thus adding carbon source is an effective way to improve denitrification and reduce N_2O emission at the same time.

The amount of organic carbon has an important effect on the N_2O emission during the biological treatment of wastewater, and the difference of the carbon source type in the sewage can also lead to different N_2O emission. Lu and Kartik (2010) used two denitrification sequencing batch reactors to study the effects of carbon source addition, nitrite accumulation and dissolved oxygen concentration (2.5, 5.1 and 9.0 mg L^{-1}, respectively), when methanol and ethanol were used as carbon sources, respectively. The N_2O emission in the two reactors was lower than 0.2% of the total nitrogen in the steady state. There was no significant difference between the N_2O emission and the N_2O emission. The N_2O emissions from the two reactors did not increase significantly during the nitrogen impingement, but there was a significant difference in the experiments of increasing the denitrification inhibition rate. The denitrification process of methanol as carbon source was restrained obviously. The removal rate of N (superscript +) and N (superscript +) were reduced obviously in one period, and the N_2O emission was only about 0.1% of the total nitrogen, but not with the dissolved oxygen. However, in the reactor with ethanol as carbon source, the denitrification rate decreased and N_2O emission rate increased sharply with the increase of dissolved oxygen concentration. When the dissolved oxygen was 9.0 mg L^{-1}, the total N of influent was 7.1% transformed to N_2O gas. The researchers believe that this phenomenon mainly due to different types of carbon sources in the reactor will be a different enrichment of denitrifying bacteria, thereby affecting the activity of denitrifying enzymes in the sludge and the adaptability to external environmental changes (Christensson et al. 1994; Hallin and Pell 1998).

It was found that glycogen-accumulating organisms (GAOs) produced a large amount of N_2O during denitrification, which would increase the risk of N_2O emission in the biological phosphorus removal process. Glycogen- and phosphate-accumulating organisms (PAOs) have similar metabolic mechanisms, absorbing organic matter in anaerobic conditions and converting them into polyhydroxybutyrate (PHB) under aerobic conditions catabolism. As the polysaccharide bacteria is not associated with anaerobic phosphate release and aerobic phosphorus uptake process, in biological phosphorus removal process, a large number of polysaccharide bacteria will seriously affect the biological phosphorus removal efficiency, but the study found that the presence of polysaccharide bacteria at high temperature is inevitable (Lopez-Vazquez et al. 2009). The reason for the high N_2O gas produced by the metabolism of endogenous carbon is that the utilization of PHB is a rate-limiting step, which leads to the denitrifying enzyme's competition for electrons. However, the affinity of N_2O reductase to electrons inhibits its activity (Murnleitner et al. 1997; Beun et al. 2002), which results in an increase in N_2O emissions.

Lemaire et al. (2006) studied the mechanism of N_2O production in a SBR reactor performing simultaneous denitrification and dephosphorization. It was found that the denitrifying bacteria in the system were mainly polysaccharides rather than polyphosphate bacteria, and more than half of the influent nitrogen was emitted in the form of N_2O. When the system was mixed with a half normal denitrification sludge and provided additional carbon source, N_2O emission immediately decreased. It was revealed that the phenomenon was mainly due to the long-term use of a single carbon source which leads to a single structure of the bacterial community in the activated sludge system. Zeng et al. (2003) succeeded in running a SBR reactor with short-range nitrification and denitrification by controlling dissolved oxygen in the aerobic zone at 0.5 mg L^{-1}. Further study showed that N_2O was mainly produced by the polysaccharide bacteria rather than the polyphosphate bacterium, which played an important role in the biosynthesis of PHA. Schalk-Otte et al. (2000) studied the N_2O production features of denitrifying bacteria *A. faecalis*. It was found that once the consumption of exogenous organic carbon turned into the endogenous carbon PHB as a substrate for growth and metabolism, N_2O production will immediately increase. Like the endogenous storage material of most bacteria, PHB plays an important role in maintaining the ecological balance of the flora in the wastewater biological treatment process, especially when the sludge environment is greatly impacted. Therefore the biological treatment process and influent water quality fluctuate greatly, and PHB metabolism may play an important role in N_2O emissions (Van Loosdrecht et al. 1997). Tabrez Khan and Hiraishi (2001) screened a denitrifying bacterium that degrades PHB. It was found that nitrification can be directly converted to nitrogen without the accumulation of intermediates during the denitrification, which is different from the previous studies. However this study is based on PHB as an external carbon source rather than direct consumption of endogenous carbon, whether there is essential difference between the need for further validation. Future studies should focus on whether N_2O production is directly related to endogenous carbon consumption, since N_2O emissions in the biological phosphorus removal system are directly related to other factors in the denitrification process, such as nitrite accumulation. Their relationship needs to be further verified experimentally.

4. pH

In biological wastewater treatment, the optimum pH value range is 6.5–8.5 and too acidic or alkaline environment is unfavorable to the survival and growth of microorganism. Also pH value has a great impact on the form and concentration of free ammonia and free nitrite, which could bring variations of N_2O emissions from the sewage treatment process. In addition, the change of pH would change the dissolution balance of N_2O between the gas and liquid phases, as the pH favors the escape of N_2O from the aqueous phase. Sumer et al. (1995) studied the effect of environmental conditions on N_2O emissions from actual wastewater treatment plants on site and found a significant correlation between pH and N_2O concentration. In the nitrification process, the pH value is widely used as an important parameter in the operation of the debugging process. In addition to its important influence on the activity of ammonia-oxidizing bacteria, the pH value also affects the concentration of free ammonia and free nitrite, as listed in 1.3 and 1.4. As the pH increases, the dissociation equilibrium moves toward the free ammonia and the free ammonia is the true substrate for ammonia-oxidizing bacteria (Suzuki et al. 1974) but inhibits the activity of nitrosomonas (Vadivelu et al. 2007). When pH is lowered, the concentration of free nitrite increases, thereby inhibiting ammonia-oxidizing bacteria and nitrite-oxidizing bacteria (Hellinga et al. 1998; Vadivelu et al. 2006).

$$NH_4^+ \leftrightarrow NH_3 + H^+ \quad pka = 9.25 \text{ at } 25^\circ C \tag{1.3}$$

$$NO_2^- + H_2O \leftrightarrow HNO_2 + OH^- \quad pka = 3.35 \text{ at } 25^{\circ}C \tag{1.4}$$

Law et al. (2011) studied the effect of pH on nitrous oxide production in partial nitrification systems and found that the lowest N_2O production rate was 0.15 ± 0.01 mg N_2O-N h^{-1} g^{-1} VSS, which was obtained at pH 6.0 and 7.0. N_2O production rate increased with increasing pH and reached the highest 0.53 ± 0.04 mg N_2O-N h^{-1} g^{-1} VSS at pH 8.0. Further study demonstrated the N_2O production rate was linear correlated with ammonia oxidation rate (AOR) but not with free ammonia or free nitrite, indicating the effect of pH on N_2O was ascribed to ammonia oxidation without the effect of free ammonia and free nitrous acid. Shiskowski and Mavinic (2006) studied the N_2O production of AOB at different pH and nitrite concentrations, which found that the change of pH mainly affected the concentration of nitrous acid that was the direct electron acceptor of AOB for autotrophic denitrification.

Hynes and Knowles (1984) studied the effect of pH on N_2O production by ammonia-oxidizing bacteria *N. europaea*. With the increase of pH from 6.0 to 8.5, the N_2O production rate increased significantly, but the researchers did not provide information on the change of nitrite salt and nitrite concentration and therefore cannot explain in depth the mechanism of pH influence on the rate of N_2O production by ammonia-oxidizing bacteria. Wrage et al. (2001) developed a theoretical model for N_2O production from autotrophic bacteria under low pH conditions based on thermodynamics. The reduction of ammonia-free energy to N_2O conversion was calculated when pH decreased, but the pH range of 7.0 to 4.0 was used by the investigators, so the variation range of most wastewater treatment processes was different and could not be applied to most of the research results.

Most studies have shown that the change of pH has important influence on N_2O production during denitrification. Hanaki et al. (1992) investigated the effect of conditional changes on the amount of N_2O produced during denitrification in a small laboratory scale, where the amount of N_2O produced at pH 6.5 was significantly higher than that at pH 7.5 and 8.5, but there were no significant differences between the latter two. Thorn and Sorensson (1996) studied the effects of environmental factors, such as carbon sources, nitrides, nitrites, and pH, on N_2O emissions in denitrification ponds. It was found that with decreasing pH, denitrification rate decreased and N_2O emissions rose and reached the maximum at pH 5.5. The emission rate was 0.31 mg N g^{-1}VSS h^{-1}, and when the pH value was higher than 6.8, no N_2O gas emission was observed.

In wastewater treatment plants, the pH range is generally between 7.0 and 8.0, which is relatively stable and does not have such a large fluctuation range as pH values in laboratory studies. So the effect of pH on N_2O emissions in practical wastewater treatment processes may be negligible compared to other environmental factors.

5. Influent Nitrogen Load

From the viewpoint of N_2O production and kinetic reaction analysis, the substrate concentration in the biochemical reaction process will directly influence the rate of nitrification and denitrification,and thus determine the level of N_2O production. With the high influent ammonia nitrogen concentration, the activity of nitrite-oxidizing bacteria was inhibited, which resulted in the accumulation of nitrite and the increase of N_2O production. In addition, the higher NH_3-N concentration in autotrophic denitrification process will provide sufficient electron donor for the autotrophic denitrification process and further promote the N_2O production in the process of autotrophic denitrification. However, if the influent contains high concentration of nitrogen, the rate of nitrate reduction will increase, because nitrate is more likely to act as an electron acceptor than N_2O, which inhibits the N_2O to nitrogen further transformation.

Lotito et al. (2012) conducted a continuous monitoring of N_2O emissions from field A/O pilot-scale reactors and found that N_2O emissions were well correlated with influent nitrogen load that was N_2O emissions increased with increasing influent nitrogen load. N_2O emissions were "inverted bell-shaped" with ammonia nitrogen and nitrite concentration appearing the peak value at the same time during the 1-day cycle in the early morning when the nitrogen concentration was high. Kim and Bae (2000) studied the gaseous products during autotrophic and heterotrophic denitrification. It was found that the N_2O production decreased with the decrease of influent nitrate concentration. When the influent nitrate concentration was 1500 mg L^{-1}, 25–40% of the nitrate was converted to N_2O in the influent, but when the nitrate concentration was reduced to 750 mg L^{-1}, there was no N_2O emission in the denitrification product, which was attributed to the need of the autotrophic denitrification process to consume substantial alkalinity. With the high nitrate concentration, the alkalinity of autotrophic denitrification process is insufficient and the process of autotrophic denitrification is not thorough enough, which leads to the increase of N_2O emission. N_2O emissions and nitrogen loading in the influent have also been shown to correlate well in a scale-activated sludge system (Sumer et al. 1995).

6. Sludge Age, Toxic Substances, Influent Mode, Salinity, and Other Factors

Sludge age is an important operating parameter in wastewater biotechnology. Because different generations of microorganisms vary in generation time, different microbial populations can be optimized by controlling the sludge age, thus affecting N_2O gas emission. Noda et al. (2003) studied the effects of sludge age on N_2O emissions from A/O activated sludge systems and found that the N_2O emission rates in the aerobic tank were 1.72, 1.00, and 0.20 μg min^{-1} for 7, 10, and 20 days, respectively. Moreover, 22 kinds of N_2O reductase gene sequences were isolated by molecular biology method, and it was considered that the dominance of this type of denitrifying bacteria in the system decreased N_2O emission during denitrification process. Hanaki et al. (1992) also show that shorter sludge ages produce more N_2O, both in the nitrification process and in the denitrification process. Lotito et al. (2012) used a pilot aeration tank to study the effects of different sludge ages on the N_2O production in the process of reduction and recovery of the dissolved oxygen. It was found that when the sludge age was 10–12 days and compare with 15 days, N_2O production was higher during the time when dissolved oxygen was reduced from 2.0 mg L^{-1} to 1.0 mg L^{-1}, and it took a longer time to reach stability after recovering to 2.0 mg L^{-1} dissolved oxygen. The reason of this phenomenon is that when the SRT is reduced, the nitrifying bacteria, especially nitrite-oxidizing bacteria, which have a long generation in the sludge system are eliminated, resulting in the incomplete nitrification of the system and the increase of the N_2O gas emission.

The presence of these toxic chemicals in wastewater treatment processes can lead to an increase in N_2O emissions due to the deleterious effects of various chemicals such as H_2S, formaldehyde, and heavy metal ions on nitrous oxide reductase activity. Schonharting et al. (1998) found that H_2S concentration higher than 0.32 mg L^{-1} would strongly inhibit the activity of anti-digestive enzymes, thereby increasing the N_2O gas emissions in biological nitrogen removal process. The researchers validate this hypothesis through the further extended denitrification model and points out that the pH value in the system indirectly influences the N_2O gas emission during the denitrification process by changing the H_2S concentration. Garrido et al. (1998) used anaerobic biofilter to study the emission of N_2O during denitrification and investigated the change of N_2O emission when formaldehyde was added. It was found that N_2O emission sharply increased to 1.0% when the formaldehyde concentration reached 1500 mg L^{-1}. The researchers believe that the concentration of formaldehyde has the detrimental effect to organisms and deduced that the concentration of N_2O in the gas phase can be used as an indicator of the normal operation of the system during the denitrification process.

Different influent modes have important influence on nitrogen removal efficiency and N_2O emission. Osada et al. (1995) compared the effects of continuous aeration (conventional process) and intermittent aeration (improved process) on N_2O emissions. It was found that 35% of total influent nitrogen was converted to N_2O gas in the continuous aeration process, and the N_2O gas emitted in the first hour after the aeration account for 60% of the total emissions. When the feedwater is changed into the step feed and intermittent aeration, the ratio of the total nitrogen into N_2O gas is less than 1%. The mechanism of this improved process to reduce N_2O emissions needs further verification, but it can be inferred from the nitrogen oxide conversions that intermittent aeration significantly reduced the accumulation of nitrite and nitrate, thereby reducing the N_2O emissions.

Domestic wastewater generally does not appear with very high salinity, but some special industrial wastewater can contain high NaCl, Na_2SO_4, and other inorganic substances causing high salinity of wastewater. Studies have shown that the high salt content in wastewater (Campos et al. 2002; Glass and Silversten 1999) severely affects nitrogen removal in biological process and leads to an increase in N_2O emissions. Tsuneda et al. (2005) studied the effects of salinity on N_2O emissions in single nitrification ponds, single denitrification ponds, and anaerobic-aerobic ponds and found that in a single nitrification tank, the N_2O conversion increased 2.2 times; but in the single denitrification process, when the salinity varied from 3.0% to 5.0%, the N_2O emission did not change significantly. In the anaerobic-aerobic activated sludge process, when the salinity increased to 3.0%, the nitrification and denitrification efficiency decreased, but then the denitrification rate was restored. At the same time, the N_2O emission in the anaerobic tank achieves stability after increasing to a certain value; the researchers believe that the reason why denitrification rate of anaerobic tank declines and the N_2O emissions increase is that sludge reflux will bring dissolved oxygen from aerobic tank to the anaerobic tank, which results in incomplete denitrification. The authors suggest that reducing the salinity below 3.0% in the anaerobic-aerobic process plays an important role in the control of N_2O emissions. However, in the normal wastewater treatment process, even in the industrial wastewater treatment process, salinity quickly increasing to more than 3.0% is also very rare, so the impact of salinity on the N_2O emissions on sewage treatment process can be basically ignored.

1.4.3 N_2O Emission Reduction in BNR Process

With the popularization and expansion of biological treatment process, the strategy of N_2O reduction is urgently needed to be effectively formulated and implemented. From the above analysis, we can see that the operating parameters that lead to the increase of N_2O emission during the sewage treatment process include the shortage of dissolved oxygen, the accumulation of nitrite, and the low ratio of carbon to nitrogen. In the course of operation, avoiding these factors will effectively reduce the N_2O gas emissions.

It is a simple and feasible way by maintaining adequate aeration in sewage biological treatment unit to reduce autotrophic denitrification and decrease the accumulation of nitrite in the system. However, excessive aeration can lead to the increase in energy consumption and also lead to inhibition of denitrification process and thereby increase N_2O emissions. Therefore the best aeration is controlled at satisfying complete nitrification without affecting denitrification. The prolongation of system sludge age was an effective means to enhance nitrification, and the increase of carton to nitrogen ratio was the guarantee of complete denitrification process. In wastewater treatment plants with low influent organic loading, the removal of organic matter in the pretreatment process should be reduced and the organic carbon source should be supplemented to meet the denitrification requirements. The biological treatment system of wastewater extremely disturbed by external environment will cause the increase of N_2O emission, such as the influent NH3-N concentration, the impact of toxic substances, and the change of pH value. Therefore, the buffer tank is set up before the biological treatment unit, and appropriately increased internal reflux and sludge reflux ratio is an effective way.

In laboratory studies, N_2O emissions from wastewater biological processes have been effectively controlled. Yang et al. (2009) found that NH_4^+ and NO_2-N concentration could be controlled to a low level and decrease N_2O emission by 50% through step-feeding process. Law et al. (2011) suggested that the rate of ammonia reduction can be effectively controlled by reducing the rate of feedwater instead of pulsed feedwater to decrease ammonia oxidation. Pellicer-Nàcher et al. (2010) successfully operated the membrane aeration biofilm reactor (MABRs) with controlled sequential aeration mode, in which N_2O production was reduced by 100 times than the general ammonia oxidation reactor due to the cooperation of aerobic and anaerobic ammonia-oxidizing bacteria in the reactor. Park et al. (2000) reduced the N_2O production in denitrification process by adding methanol to reduce the competition of denitrifying enzymes for electrons. Fukumoto et al. (2006) reduced the N_2O emission significantly by reducing the accumulation of nitrite by adding nitrite-oxidizing bacteria to the sludge system. Park et al. (2007) reduced N_2O emissions by using the immobilized aerobic denitrifying bacteria *Alcaligenes faecalis*. Takaya et al. (2003) screened two aerobic denitrifying bacteria, P. stutzeri TR2 and Pseudomonas sp. 50, and found that only two trace amounts of N_2O were produced during aerobic denitrification, indicating a potential to reduce N_2O emission in wastewater treatment. Ikeda-Ohtsubo et al. (2013) inoculated the strain *P. stutzeri* TR2 into the pilot SBR process, which could be present for more than 32 days at 40–45 °C and 2–5 mM nitrite and reduced N_2O emission from the activated sludge system.

However, it is worth noting that the N_2O control method in wastewater biological treatment is only widely studied in the laboratory, but has not been validated and applied in actual sewage treatment plants, so how to carry out in the actual sewage treatment process in the future N_2O emission control strategy remains to be further studied and improved.

1.5 The Content and Significance of this Study

1.5.1 Research Content

The traditional biological denitrification theory is based on the nitrification and denitrification of microorganisms, that is, under the aerobic condition, the ammonia nitrogen is oxidized by the autotrophic bacteria to nitrogen nitrite and nitrogen and then reduced by the heterotrophic bacteria to gaseous state under the hypoxic condition to achieve the purpose of denitrification wastewater. Therefore, in the traditional denitrification process, microorganisms need to continue to undergo alternating aerobic and anoxic environment, which will inevitably lead to dissolved oxygen into the denitrification process, thereby inhibiting denitrifying enzymes, especially the activity of N_2O reductase, resulting in a large number of N_2O emissions. Therefore, how to effectively control the N_2O emission during the process of wastewater denitrification has been paid more and more attention by the researchers.

Aerobic denitrification is a new biological denitrification technology proposed in 1980s, which is based on aerobic denitrification bacteria that can simultaneously carry out the characteristics of oxygen and nitrate respiration, providing the theoretical feasibility of denitrification in a single aerobic reactor. Since the first aerobic denitrifying bacterium *T. pantotropha* was successfully isolated, and more and more aerobic denitrifying strains were isolated in recent years. However, due to the sensitivity of N_2O reductase to O_2, most strains end with N_2O as denitrification product instead of N_2, which greatly reduced the application potential of aerobic denitrifying bacteria. Therefore, it is of great theoretical and practical value to select high-efficient aerobic denitrifying bacteria which can rapidly utilize N_2O and achieve nitrogen oxide emission reduction in practical wastewater treatment.

The main contents of this study include the following aspects:

1. Optimization of N_2O efficient aerobic denitrification and its characterization

Using the method of isotopic labeling and denitrification monitoring, the strain of N_2O was quickly selected from 14 strains of aerobic denitrifying bacteria and identified by 16S rRNA sequence. The effects of dissolved oxygen, pH, and C/N ratio on the denitrification process were investigated under the conditions of ammonia nitrogen, nitrogen nitrite, nitrite, and N_2O as different nitrogen sources.

2. Study on the rapid utilization of nitrogen oxides in aerobic denitrification

Quantitative analysis of functional gene expression during aerobic denitrification was performed by reverse transcription-fluorescence quantitative real-time PCR. Combined with the change of metabolite concentration during denitrification, the mechanisms of aerobic denitrification and nitrogen oxides utilization were explained.

3. Study on the emission characteristics and mechanism of nitrogen oxides in activated sludge system

The emission factors and time-space characteristics of nitrogen oxides in stable operation and environmental disturbances were investigated by means of denitrifying SBR and pilot-scale Carrousel oxidation ditch. The contribution of

N_2O emission by autotrophic denitrification and heterotrophic denitrification to the total N_2O emission was studied with the help of batch experiments, and the mechanism of N_2O production was further interpreted by the microbial community structure.

4. Applied research on nitrogen oxide reduction in activated sludge system

Based on the unique advantages of rapid utilization of N_2O in aerobic denitrification and the emission characteristics of nitrogen oxides in SBR and oxidation ditch systems, the effects of bioaugmentation on nitrogen oxides emission reduction in activated sludge systems were investigated, which provided experimental and theoretical basis for the engineering application.

5. Application of strains in the biofilm system to control nitrogen oxide emissions

Based on the unique advantages of the rapid reduction of NO by aerobic denitrification, the selected bacterial strain was applied to construct a trickling filter reactor for flue gas purification under different concentration of oxygen and nitrogen oxides. The results demonstrated an excellent NO removal efficiency, which provided a basis for the future application of aerobic denitrifying bacteria for flue gas purification.

The technical roadmap for this study is shown in Fig. 1.4.

1.5.2 Innovation Points

The innovation of this study is mainly reflected in the following aspects:

1. Two efficient aerobic denitrifying strains which can rapidly utilize nitrogen oxides in aerobic denitrification process are provided, demonstrating a new way to solve the bottlenecks encountered in practical application of aerobic denitrifying bacteria.
2. Through the quantitative analysis of gaseous products and functional gene expression in the process of denitrification, the mechanism of rapid utilization of nitrogen oxides in aerobic denitrification was revealed.
3. Through bioaugmentation experiments, the feasibility of reducing nitrogen oxides emission in activated sludge system and biofilm system was determined, which laid an experimental and theoretical basis for engineering application.

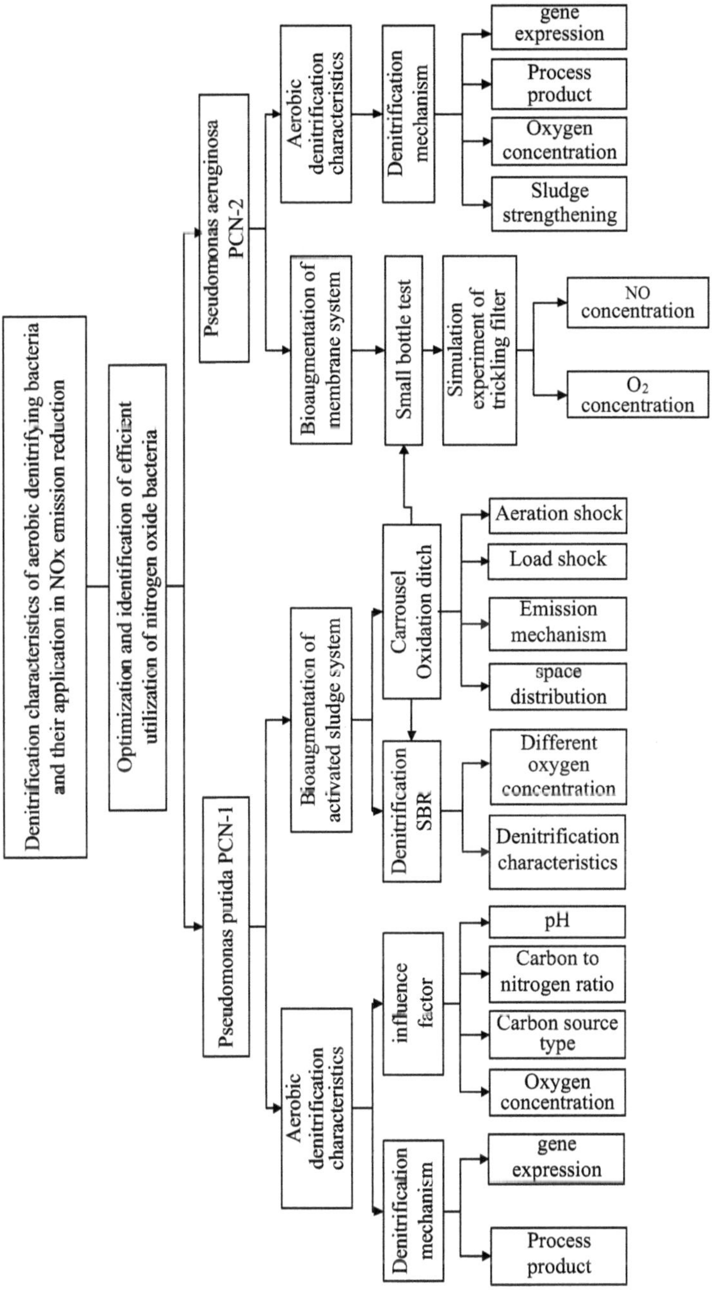

Fig. 1.4 Technology roadmap of this research

References

Aboobakar, A., Cartmell, E., Stephenson, T., Jones, M., Vale, P., & Dotro, G. (2013). Nitrous oxide emissions and dissolved oxygen profiling in a full-scale nitrifying activated sludge treatment plant. *Water Research, 47*, 524–534.

Ahn, J. H., Kim, S. P., Park, H. K., Rahm, B., Pagilla, K., & Chandran, K. (2010). N_2O Emissions from activated sludge processes, 2008–2009: Results of a national monitoring survey in the United States. *Environmental Science and Technology, 44*, 4505–4511.

Anderson, I. C., Poth, M., Homstead, J., & Burdige, D. (1993). A comparison of nitric oxide and nitrous oxide production by the autotrophic nitrifier *Nitrosomonas europaea* and the heterotrophic nitrifier *Alcaligenes faecalis*. *Applied and Environmental Microbiology, 59*, 3525–3533.

Arts, P. A. M., Robertson, L. A., & Kuenen, J. G. (1995). Nitrification and denitrification by Thisphaera pantotropha in aerobic chemostat cultures. *FEMS Microbiology Ecology, 18*, 305–316.

Barnes, J. M., Apel, W. A., & Barrett, K. B. (1995). Removal of nitrogen oxides from gas streams using biofiltration. *Journal of Hazardous Materials, 41*, 315–326.

Beaumont, H., Lens, S., Reijinders, W., Westerhoff, H., & van Spanning, R. (2004). Expression of nitrite reductase in *Nitrosomonas europaea* involves NsrR, a novel nitrite-sensitive transcription repressor. *Molecular Microbiology, 54*, 148–158.

Bell, L. C., Richardson, D. J., & Ferguson, S. J. (1990). Periplasmic and membrane-bound respiratory nitrate reductases in *Thiosphaera pantotropha*. The periplasmic enzyme catalyzes the first step in aerobic denitrification. *FEBS Letter, 4*, 85–87.

Bergaust, L., Shapleigh, J., Frostegard, A., & Bakken, L. (2008). Transcription and activities of NO_x reductases in Agrobacterium tumefaciens: The influence of nitrate, nitrite and oxygen availability. *Environmental Microbiology, 10*, 3070–3081.

Berks, B. C., Baratta, D., Richardson, J., & Ferguson, S. J. (1993). Purification and characterization of a nitrous oxide reductase from *Thiosphaera pantotropha*. Implications for the mechanism of aerobic nitrous oxide reduction. *European Journal of Biochemistry, 212*, 467–476.

Beun, J. J., Dircks, K., Van Loosdrecht, M. C., & Heijnen, J. J. (2002). Poly-beta-hydroxybutyrate metabolism in dynamically fed mixed microbial cultures. *Water Research, 36*, 1167–1180.

Bock, E., Schmidt, I., Stuven, R., & Zart, D. (1995). Nitrogen loss caused by denitrifying *Nitrosomonas* cells using ammonium or hydrogen as electron donors and nitrite as electron acceptor. *Archives of Microbiology, 163*, 16–20.

Braker, G., Zhou, J., Wu, L., Devol, A. H., & Tiedje, J. M. (2000). Nitrite reductase genes (*nirK* and *nirS*) as functional markers to investigate diversity of denitrifying bacteria in pacific northwest marine sediment communities. *Applied and Environmental Microbiology, 66*, 2096–2104.

Butler, M. D., Wang, Y. Y., Cartmell, E., & Stephenson, T. (2009). Nitrous oxide emissions for early warning of biological nitrification failure in activated sludge. *Water Research, 43*, 1265–1272.

Campos, J. L., Mosquera-Corral, A., Sanchez, M., Mendez, R., & Lema, J. M. (2002). Nitrification in saline wastewater with high ammonia concentration in an activated sludge unit. *Water Research, 36*, 2555–2560.

Cantera, J. J. L., & Stein, L. Y. (2007). Molecular diversity of nitrite reductase genes (*nirK*) in nitrifying bacteria. *Environmental Microbiology, 9*, 765–776.

Casciotti, K. L., & Ward, B. B. (2001). Dissimilatory nitrite reductase genes from autotrophic ammonia-oxidizing bacteria. *Applied and Environmental Microbiology, 67*, 2213–2221.

Chen, Q., & Ni, J. R. (2011). Heterotrophic nitrification-aerobic denitrification by novel isolated bacteria. *Journal of Industrial Microbiology and Biotechnology, 38*, 1305–1310.

Chen, P. Z., Li, J., Li, Q. X., Wang, Y. C., Li, S. P., Ren, T. Z., & Wang, L. G. (2012). Simultaneous heterotrophic nitrification and aerobic denitrification by bacterium Rhodococcus sp. CPZ24. *Bioresource Technology, 116*, 266–270.

Chen, J. W., & Strous, M. (2013). Denitrification and aerobic respiration, hybrid electron transport chains and co-evolution. *Biochimica et Biophysica Acta, 1827*, 136–144.

Chen, M., Wang, W., Feng, Y., Zhu, X., Zhou, H., Tan, Z., & Li, X. (2014). Impact resistance of different factors on ammonia removal by heterotrophic nitrificationaerobic denitrification bacterium Aeromonas sp. HN-02. *Bioresource Technology, 167*, 456–461.

Chen, D., Yang, K., Wang, H., Lv, B., & Ma, F. (2015). Characteristics of nitrate removal in a bio-ceramsite reactor by aerobic denitrification. *Environmental Technology, 36*, 1457–1463.

Christensson, M., Lie, E., & Welander, T. (1994). A comparison between ethanol and methanol as carbon sources for denitrification. *Water Science and Technology, 30*, 83–90.

Chuang, H. P., Ohashi, A., Imachi, H., Tandukar, M., & Harada, H. (2007). Effective partial nitrification to nitrite by down-flow hanging sponge reactor under limited oxygen condition. *Water Research, 41*, 295–302.

Czepiel, P., Crill, P., & Harriss, R. (1995). Nitrous oxide emissions from municipal wastewater treatment. *Environmental Science and Technology, 29*, 2352–2356.

Daigger, G. T., & Littleton, H. X. (2000). Characterization of simultaneous nutrient removal in staged, closed-loop bioreactors. *Water Environmental Research, 72*, 330.

Dalal Ram, C., Robertson, W. W., Philip, G., & Parton, W. J. (2003). Nitrous oxide emission from Australian agricultural lands and mitigation options: A review. *Australian Journal of Chemistry, 41*, 165–195.

Dean, J. A. (1992). *Lange's handbook of chemistry*. New York: McGraw-Hill, Inc..

Desloover, J., De Clippeleir, H., Boeckx, P., Du Laing, G., Colsen, J., Verstraete, W., & Vlaeminck, S. E. (2011). Floc-based sequential partial nitration and anammox at full scale with contrasting N_2O emissions. *Water Research, 45*, 2811–2821.

Duan, J., Fang, H., Su, B., Chen, J., & Lin, J. (2015). Characterization of a halophilic heterotrophic nitrification-aerobic denitrification bacterium and its application on treatment of saline wastewater. *Bioresource Technology, 179*, 421–428.

Foley, J., De Haas, D., Yuan, Z., & Lant, P. (2010). Nitrous oxide generation in full scale BNR wastewater treatment plants. *Water Research, 44*, 831–844.

Frette, L., Gejlsbjerg, B., & Westermann, P. (1997). Aerobic denitrifiers isolated from an alternating activated sludge system. *FEMS Microbiology Ecology, 24*, 363–370.

Fujiwara, T., & Fukumori, Y. (1996). Cytochrome *cb*-type nitric oxide reductase with cytochrome *c* oxidase activity from *Paracoccus denitrificans* ATCC 35512. *Journal of Bacteriology, 178*, 1866–1871.

Fukumoto, Y., Suzuki, K., Osada, T., Kuroda, K., Hanajima, D., Yasuda, T., & Haga, K. (2006). Reduction of nitrous oxide emission from pig manure composting by addition of nitrite-oxidizing bacteria. *Environmental Science and Technology, 40*, 6787–6791.

Fulop, V., Moir, J. W., Ferguson, S. J., & Hajdu, J. (1995). The anatomy of a bifunctional enzyme: Structural basis for reduction of oxygen to water and synthesis of nitric oxide by cytochrome *cd1*. *Cell, 81*, 369–377.

Garbeva, P., Baggs, E. M., & Prosser, J. I. (2007). Phylogeny of nitrite reductase (*nirK*) and nitric oxide reductase (*norB*) genes from *Nitrosospira* species isolated from soil. *FEMS Microbiology Letter, 266*, 83–89.

Garrido, J. M., Moreno, J., Mendez-Pampin, R., & Lema, J. M. (1998). Nitrous oxide production under toxic conditions in a denitrifying anoxic filter. *Water Research, 32*, 2550–2552.

Glass, C., & Silverstein, J. (1999). Denitrification of high-nitrate, high salinity wastewater. *Water Research, 33*, 223–229.

Goreau, T. J., Kaplan, W. A., Wofsy, S. C., McElroy, M. B., Valois, F. W., & Watson, S. W. (1980). Production of NO_2 and N_2O by nitrifying bacteria at reduced concentrations of oxygen. *Applied and Environmental Microbiology, 40*, 526–532.

Guo, L., Lamaire Chad, C., Bellandi, G., Daelman, M., Amerlinck, Y., Maere, T., Nous, J., Flameling, T., Weijers, S., van Loosdrecht, M. C. M., Volcke, E. I. P., Nopens, I., & Vanrolleghem, P. A. (2013). *High-frequency field measurement of nitrous oxide (N_2O) gas emissions and influencing factors at WWTPs under dry and wet weather conditions*. Conference: WEF/IWA Nutrient Removal and Recovery, July 29–31, 2013, At Vancouver, Canada.

Gupta, A. B., & Gupta, S. K. (2001). Simultaneous carbon and nitrogen removal in a mixed culture aerobic RBC biofilm. *Water Research, 35*, 1714–1722.

Hallin, S., & Pell, M. (1998). Metabolic properties of denitrifying bacteria adapting to methanol and ethanol in activated sludge. *Water Research, 32*, 13–18.

Hanaki, K., Hong, Z., & Mstsuo, T. (1992). Production of nitrous oxide gas during denitrification of wastewater. *Water Science and Technology, 26*, 1027–1036.

Hellinga, C., Schellen, A. A. J. C., Mulder, J. W., van Loosdrecht, M. C. M., & Heijnen, J. J. (1998). The SHARON process: An innovative method for nitrogen removal from ammonium-rich waste water. *Water Science and Technology, 37*, 135–142.

Hu, Z., Zhang, J., Xie, H., Li, S., Wang, H., & Zhang, T. (2011). Effect of anoxic/aerobic phase fraction on N_2O emission in a sequencing batch reactor under low temperature. *Bioresource Technology, 102*, 5486–5491.

Huang, X. F., Li, W. G., Zhang, D. Y., & Qin, W. (2013). Ammonium removal by a novel oligotrophic *Acinetobacter sp.* Y16 capable of heterotrophic nitrification–aerobic denitrification at low temperature. *Bioresource Technology, 146*, 44–50.

Hynes, R. K., & Knowles, R. (1984). Production of nitrous oxide by Nitrosomonas europaea: Effects of acetylene, pH, and oxygen. *Canadian Journal of Microbiology, 30*, 1397–1404.

Ikeda-Ohtsubo, W., Miyahara, M., Kim, S. W., Yamada, T., Matsuoka, M., Watanabe, A., Fushinobu, S., Wakagi, T., Shoun, H., Miyauchi, K., & Endo, G. (2013). Bioaugmentation of a wastewater bioreactor system with the nitrous oxide-reducing denitrifier *Pseudomonas stutzeri* strain TR2. *Journal of Bioscience and Bioengineering, 115*, 37–42.

IPCC. (2007). Climate change 2007: Synthesis report. In Core Writing Team, R. K. Pachauri, & A. Reisinger (Eds.), *Contribution of Working groups I, II and III to the fourth assessment report of the intergovernmental panel on climate change*. Geneva: IPCC.

Itokawa, H., Hanaki, K., & Matsuo, T. (2001). Nitrous oxide production in high-loading biological nitrogen removal process under low COD/N ratio condition. *Water Research, 35*, 657–664.

Jia, W., Liang, S., Zhang, J., Ngo, H. H., Guo, W., Yan, Y., & Zou, Y. (2013). Nitrous oxide emission in low-oxygen simultaneous nitrification and denitrification process: Sources and mechanisms. *Bioresource Technology, 136*, 444–451.

Joo, H. S., Hirai, M., & Shoda, M. (2006). Piggery wastewater treatment using Alcaligenes faecalis No. 4 with heterotrophic nitrification and aerobic denitrification. *Water Research, 40*, 3029–3026.

Joss, A., Salzgeber, D., Eugster, J., & Konig, R. (2009). Full-scale nitrogen removal from digester liquid with partial nitration and anammox in one SBR. *Environmental Science and Technology, 43*, 5301–5306.

Kampschreur, M. J., Tan, N. C., Kleerebezem, R., Picioreanu, C., Jetten, M. S. M., & Van Loosdrecht, M. C. M. (2008a). Effect of dynamic process conditions on nitrogen oxide emission from a nitrifying culture. *Environmental Science and Technology, 42*, 429–435.

Kampschreur, M. J., van der Star, W. R. L., Wielders, H. A., Mulder, J. W., Jetten, M. S. M., & van Loosdrecht, M. C. M. (2008b). Dynamics of nitric oxide and nitrous oxide emission during full-scale reject water treatment. *Water Research, 42*, 812–826.

Kampschreur, M. J., Poldermans, R., Kleerebezem, R., van der Star, W. R., Haarhuis, R., Abma, W. R., Jetten, M. S., Jetten, M. S., & van Loosdrecht, M. C. (2009). Emission of nitrous oxide and nitric oxide from a full-scale single-stage nitritation-anammox reactor. *Water Science and Technology, 60*, 3211–3217.

Kester, R. A., De Boer, W., & Laanbroek, H. J. (1997). Production of NO and N_2O by pure cultures of nitrifying and denitrifying bacteria during changes in aeration. *Applied and Environmental Microbiology, 63*, 3872–3877.

Kim, E. W., & Bae, J. H. (2000). Alkalinity requirements and the possibility of simultaneous heterotrophic denitrification during sulfurutilizing autotrophic denitrification. *Water Science and Technology, 42*, 233–238.

Kim, S. W., Miyahara, M., Fushinobu, S., Wakagi, T., & Shoun, H. (2010). Nitrous oxide emission from nitrifying activated sludge dependent on denitrification by ammonia-oxidizing bacteria. *Bioresource Technology, 101*, 3958–3963.

Kimochi, Y., Inamori, Y., Mizuochi, M., Xu, K. Q., & Matsumura, M. (1998). Nitrogen removal and N_2O emission in a full-scale domestic wastewater treatment plant with intermittent aeration. *Journal of Fermentation and Bioengineering, 86*, 202–206.

Konneke, M., Bernhard, A. E., de la Torre, J. R., Walker Christopher, B., Waterbury, J. B., & Stahl, D. A. (2005). Isolation of an autotrophic ammonia-oxidizing marine archaeon. *Nature, 437*, 543–546.

Korner, H., & Zumft, W. G. (1989). Expression of denitrification enzymes in response to the dissolved oxygen level and respiratory substrate in continuous culture of *Pseudomonas stutzeri. Applied and Environment Microbiology, 55*, 1670–1676.

Kshirsagar, M., Gupta, A. B., & Gupta, S. K. (1995). Aerobic denitrification studies on activated sludge mixed with *Thiosphaera pantotropha. Environmental Technology, 16*, 35–43.

Kundu, P., Pramanik, A., Dasgupta, A., Mukherjee, S., & Mukherjee, J. (2014). Simultaneous heterotrophic nitrification and aerobic denitrification by Chryseobacterium sp. R31 isolated from abattoir wastewater. *BioMed Research International, 4*, 436056.

Law, Y., Lant, P., & Yuan, Z. (2011). The effect of pH on N_2O production under aerobic conditions in a partial nitritation system. *Water Research, 45*, 5934–5944.

Law, Y., Ni, B. J., Lant, P., & Yuan, Z. (2012). N_2O production rate of an enriched ammonia-oxidising bacteria culture exponentially correlates to its ammonia oxidation rate. *Water Research, 46*, 3409–3419.

Lemaire, R., Meyer, R., Taske, A., Crocetti, G. R., Keller, J., & Yuan, Z. (2006). Identifying causes for N_2O accumulation in a lab-scale sequencing batch reactor performing simultaneous nitrification, denitrification and phosphorus removal. *Journal of Biotechnology, 122*, 62–72.

Liang, H., Yang, J., & Gao, D. (2014). N_2O emission from nitrogen removal via nitrite in oxic-anoxic granular sludge sequencing batch reactor. *Jounal of Environmental Science, 26*, 537–541.

Lloyd, D. (1993). Aerobic denitrification in soils and sediments - from fallacies to facts. *Trends in Ecology and Evolution, 8*, 352–356.

Lopez-Vazquez, C. M., Oehmen, A., Hooijmans, C. M., Brdjanovic, D., Gijzen, H. J., Yuan, Z., & van Loosdrecht, M. C. M. (2009). Modeling the PAO-GAO competition: Effects of carbon source, pH and temperature. *Water Research, 43*, 450–462.

Lotito, A. M., Wunderlin, P., Joss, A., Kipf, M., & Siegrist, H. (2012). Nitrous oxide emissions from the oxidation tank of a pilot activated sludge plant. *Water Research, 46*, 3563–3573.

Lu, H. J., & Kartik, C. (2010). Factors promoting emissions of nitrous oxide and nitric oxide from denitrifying sequencing batch reactors operated with methanol and ethanol as electron donors. *Biotechnology and Bioengineering, 106*, 390–3981.

Miyahara, M., Kim, S. W., Fushinobu, S., Takaki, K., Yamada, T., Watanabe, A., Miyauchi, K., & Endo, G. (2010). Potential of aerobic denitrification by Pseudomonas stutzeri TR2 to reduce nitrous oxide emissions from wastewater treatment plants. *Applied and Environmental Microbiology, 76*, 4619–4625.

Moir, J. W., Baratta, D., Richardson, D. J., & Ferguson, S. J. (1993). The purification of a cd1-type nitrite reductase from, and the absence of a copper-type nitrite reductase from, the aerobic denitrifier *Thiosphaera pantotropha*; the role of pseudoazurin as an electron donor. *European Journal of Biochemistry, 212*, 377–385.

Mulder, A., van de Graaf, A. A., Robertson, L. A., & Kuenen, J. G. (1995). Anaerobic ammonium oxidation discovered in a denitrifying fluidized bed reactor. *FEMS Microbiology Ecology, 16*, 177–183.

Murnleitner, E., Kuba, T., Van Loosdrecht, M. C. M., & Heijnen, J. J. (1997). An integrated metabolic model for the aerobic and denitrifying biological phosphorus removal. *Biotechnology and Bioengineering, 54*, 434–450.

Noda, N., Kaneko, N., Mikami, M., Kimochi, Y., Tsuneda, S., Hirata, A., Mizuochi, M., & Inamori, Y. (2003). Effects of SRT and DO on N_2O reductase activity in an anoxic-oxic activated sludge system. *Water Science and Technology, 48*, 363–370.

Osada, T., Kuroda, K., & Yonaga, M. (1995). Reducing nitrous oxide gas emissions from fill-and-draw type activated sludge process. *Water Research, 29*, 1607–1608.

Otte, S., Grobben, N. G., Robertson, L. A., Jetten, M. S., & Kuenen, J. G. (1996). Nitrous oxide production by *Alcaligenes faecalis* under transient and dynamic aerobic and anaerobic conditions. *Applied and Environmental Microbiology, 62*, 2421–2426.

Papen, H., von Berg, R., Hinkel, I., Thoene, B., & Rennenberg, H. (1989). Heterotrophic nitrification by Alcaligenes faecalis: NO_{2-}, NO_{3-}, N_2O, and NO production in exponentially growing cultures. *Applied and Environmental Microbiology, 55*, 2068–2072.

Park, K. Y., Inamori, Y., Mizuochi, M., & Ahn, K. H. (2000). Emission and control of nitrous oxide from a biological wastewater treatment system with intermittent aeration. *Journal of Bioscience and Bioengineering, 90*, 247–252.

Park, H. D., Wells, G. F., Bae, H., Criddle, C. S., & Francis, C. A. (2006). Occurrence of ammonia-oxidizing Archaea in wastewater treatment plant bioreactors. *Applied and Environmental Microbiology, 72*, 5643–5647.

Park, K. Y., Kim, S. J., Jung, J. Y., & Lee, S. H. (2007). Reduction of N_2O emission from biological nitrogen removal processes by Alcaligenes faecalis augmentation. *Journal of Industrial and Engineering Chemistry, 13*, 508–511.

Patureau, D., Bernet, N., Delgenes, J. P., & Moletta, R. (2000). Effect of dissolved oxygen and carbon nitrogen loads on denitrification by an aerobic consortium. *Applied Microbiology and Biotechnology, 54*, 535–542.

Pellicer-Nàcher, C., Sun, S., Lackner, S., Terada, A., Schreiber, F., Zhou, Q., & Smets, B. F. (2010). Sequential aeration of membrane-aerated biofilm reactors for high-rate autotrophic nitrogen removal: Experimental demonstration. *Environmental Science and Technology, 44*, 7628–7634.

Philippot, L., Mirleau, P., Mazurier, S., Siblot, S., Hartmann, A., Lemanceau, P., & Germon, J. C. (2001). Characterization and transcriptional analysis of Pseudomonas fluorescens denitrifying clusters containing the *nar, nir, nor* and *nos* genes. *Biochimica et Biophysica Acta, 1517*, 436–440.

Pijuan, M., Torà, J., Rodríguez-Caballero, A., César, E., Carrera, J., & Pérez, J. (2014). Effect of process parameters and operational mode on nitrous oxide emissions from a nitritation reactor treating reject wastewater. *Water Research, 49*, 23–33.

Poth, M., & Focht, D. (1985). 15N Kinetic analysis of N_2O production by Nitrosomonas europaea: An examination of nitrifier denitrification. *Applied and Environmental Microbiology, 49*, 1134–1141.

Poughon, L., Dussap, C. G., & Gros, J. B. (2001). Energy model and metabolic flux analysis for autotrophic nitrifiers. *Biotechnology and Bioengineering, 72*, 416–433.

Quan, X., Zhang, M., Lawlor, P. G., Yang, Z., & Zhan, X. (2012). Nitrous oxide emission and nutrient removal in aerobic granular sludge sequencing batch reactors. *Water Research, 46*, 4981–4990.

Ravishankara, A. R., Daniel, J. S., & Portmann, R. W. (2009). Nitrous oxide (N_2O): The dominant ozone-depleting substance emitted in the 21st century. *Science, 326*, 123–125.

Robertson, L. A., & Kuenen, J. G. (1984). Aerobic denitrification: A controversy revived. *Archives of Microbiology, 139*, 351–354.

Robertson, L. A., Van Niel, E. W. J., Torremans, T. A. M., & Kuenen, J. G. (1988). Simultaneous nitrification and denitrification in aerobic chemostat cultures of Thiosphaera pantotropha. *Applied and Environmental Microbiology, 54*, 2812–2818.

Santoro, A. E., Buchwald, C., Mcllvin, M. R., & Casciotti, K. (2011). Isotopic signature of N_2O produced by marine ammonia-oxidizing archaea. *Science, 333*, 1282–1285.

Schalk-Otte, S., Seviour, R. J., Kuenen, J. G., & Jetten, M. S. M. (2000). Nitrous oxide (N_2O) production by *Alcaligenes faecalis* during feast and famine regimes. *Water Research, 34*, 2080–2088.

Schmidt, I., Sliekers, O., Schmid, M., Bock, E., Fuerst, J., Kuenen, J. G., Jetten, M. S., & Strous, M. (2003). New concepts of microbial treatment processes for the nitrogen removal in wastewater. *FEMS Microbiology Reviews, 27*, 481–492.

Schonharting, B., Rehner, R., Metzger, J. W., Krauth, K., & Rizzi, M. (1998). Release of nitrous oxide (N_2O) from denitrifying activated sludge caused by H2S-containing wastewater: Quantification and application of a new mathematical model. *Water Science and Technology, 38*, 237–246.

Schulthess, R. V., Wild, D., & Gujer, W. (1994). Nitric and nitrous oxides from denitrifying activated sludge at low oxygen concentration. *Water Science and Technology, 30*, 123–132.

Sears, H. J., Sawers, G., Berks, B. C., Ferguson, S. J., & Richardson, D. J. (2000). Control of periplasmic nitrate reductase gene expression (*napEDABC*) from *Paracoccus pantotrophus* in response to oxygen and carbon substrates. *Microbiology, 146*, 2977–2985.

Shiskowski, D. M., & Mavinic, D. S. (2006). The influence of nitrite and pH (nitrous acid) on aerobic-phase, autotrophic N_2O generation in a wastewater treatment bioreactor. *Journal of Environmental Engineering Science, 5*, 273–283.

Sommer, J., Ciplak, A., Sumer, E., Benckiser, G., & Ottow, J. C. G. (1998). Quantification of emitted and retained N_2O in a municipal wastewater treatment plant with activated sludge and nitrification-denitrification units. *Agrobiological Research, 51*, 59–73.

Sumer, E., Weiske, A., Benckiser, G., & Ottow, J. C. G. (1995). Influence of environmental conditions on the amount of N_2O released from activated sludge in a domestic waste water treatment plant. *Cellular and Molecular Life Sciences, 51*, 419–422.

Sun, S., Cheng, X., & Sun, D. (2013). Emission of N_2O from a full-scale sequencing batch reactor wastewater treatment plant: Characteristics and influencing factors. *International Biodeterioration & Biodegradation, 85*, 545–549.

Sun, Y., Li, A., Zhang, X., & Ma, F. (2015). Regulation of dissolved oxygen from accumulated nitrite during the heterotrophic nitrification and aerobic denitrification of Pseudomonas stutzeri T13. *Applied Microbiology and Biotechnology, 99*, 3243–3248.

Sutka, R. L., Ostrom, N. E., Ostrom, P. H., Breznak, J. A., Gandhi, H., Pitt, A. J., & Li, F. (2006). Distinguishing nitrous oxide production from nitrification and denitrification on the basis of isotopomer abundances. *Applied and Environmental Microbiology, 72*, 638–644.

Suzuki, I., Dular, U., & Kwok, S. C. (1974). Ammonia or ammonium ion as substrate for oxidation by Nitrosomonas europaea cells and extracts. *Journal of Bacteriology, 120*, 556–558.

Tabrez Khan, S., & Hiraishi, A. (2001). Isolation and characterization of a new poly (3-hydroxybutyrate)-degrading, denitrifying bacterium from activated sludge. *FEMS Microbiology Letters, 205*, 253–257.

Takaya, N., Catalan-Sakairi, M. A., Sakaguchi, Y., Kato, I., Zhou, Z., & Shoun, H. (2003). Aerobic denitrifying bacteria that produce low levels of nitrous oxide. *Applied and Environmental Microbiology, 69*, 3152–3157.

Tallec, G., Garnier, J., Billen, G., & Gousailles, M. (2006). Nitrous oxide emissions from secondary activated sludge in nitrifying conditions of urban wastewater treatment plants: Effect of oxygenation level. *Water Research, 40*, 2972–2980.

Tallec, G., Garnier, J., Billen, G., & Gousailles, M. (2008). Nitrous oxide emissions from denitrifying activated sludge of urban wastewater treatment plants, under anoxia and low oxygenation. *Bioresource Technology, 99*, 2200–2209.

Thorn, M., & Sorensson, F. (1996). Variation of nitrous oxide formation in the denitrification basin in a wastewater treatment plant with nitrogen removal. *Water Research, 30*, 1543–1547.

Tsuneda, S., Mikami, M., Kimochi, Y., & Hirata, A. (2005). Effect of salinity on nitrous oxide emission in the biological nitrogen removal process for industrial wastewater. *Journal of Hazardous Materials, 119*, 93–98.

Vadivelu, V. M., Yuan, Z., Fux, C., & Keller, J. (2006). The inhibitory effects of free nitrous acid on the energy generation and growth processes of an enriched Nitrobacter culture. *Environmental Science and Technology, 40*, 4442–4448.

Vadivelu, V. M., Keller, J., & Yuan, Z. (2007). Effect of free ammonia on the respiration and growth processes of an enriched Nitrobacter culture. *Water Research, 41*, 826–834.

Van Cleemput, O. (1998). Subsoils: Chemo- and biological denitrification, N_2O and N_2 emissions. *Nutrient Cycling in Agroecosystems, 52*, 187–194.

Van Loosdrecht, M. C. M., Pot, M. A., & Heijnen, J. J. (1997). Importance of bacterial storage polymers in bioprocesses. *Water Science and Technology, 35*, 41–47.

Van Niel, E. W. J., Arts, P. A. M., Wesselink, B. J., Robertson, L. A., & Kuenen, J. G. (1993). Competition between heterotrophic and autotrophic nitrifiers for ammonia in chemostat cultures. *FEMS Microbiology Ecology, 102*, 109–118.

Wan, C. L., Yang, X., Lee, D. J., Du, M. A., Wan, F., & Chen, C. (2011). Aerobic denitrification by novel isolated strain using NO_2^--N as nitrogen source. *Bioresource Technology, 102*, 7244–7248.

Wang, P., Li, X., Xiang, M., & Zhai, Q. (2007). Characterization of efficient aerobic denitrifiers isolated from two different sequencing batch reactors by 16S-rRNA analysis. *Journal of Bioscience and Bioengineering, 103*, 563–567.

Wrage, N., Velthol, G. L., van Beusichem, M. L., & Oenema, O. (2001). Role of nitrifier denitrificaiton in the production of nitrous oxide. Role of nitrifier denitrificaiton in the production of nitrous oxide. *Soil Biology and Biochemistry, 33*, 1723–1732.

Wunderlin, P., Mohn, J., Joss, A., Emmenegger, L., & Siegrist, H. (2012). Mechanisms of N_2O production in biological wastewater treatment under nitrifying and denitrifying conditions. *Water Research, 46*, 1027–1037.

Yang, Q., Liu, X., Peng, C., Wang, S., Sun, H., & Peng, Y. (2009). N_2O production during nitrogen removal via nitrite from domestic wastewater: Main sources and control method. *Environmental Science and Technology, 43*, 9400–9406.

Yang, X. P., Wang, S. M., Zhang, D. W., & Zhou, L. X. (2011). Isolation and nitrogen removal characteristics of an aerobic heterotrophic nitrifying-denitrifying bacterium, *Bacillus subtilis* A1. *Bioresource Technology, 102*, 854–862.

Yao, S., Ni, J. R., Ma, T., & Li, C. (2013). Heterotrophic nitrification and aerobic denitrification at low temperature by a newly isolated bacterium, Acinetobacter sp. HA2. *Bioresource Technology, 139*, 80–86.

Ye, L., Ni, B. J., Law, Y., Byers, C., & Yuan, Z. (2014). A novel methodology to quantify nitrous oxide emissions from full-scale wastewater treatment systems with surface aerators. *Water Research., 48*, 257–268.

Yu, R., Kampschreur, M. J., Loosdrecht, M. C. M. V., & Chandran, K. (2010). Mechanisms and specific directionality of autotrophic nitrous oxide and nitric oxide generation during transient anoxia. *Environmental Science and Technology, 44*, 1313–1319.

Zeng, R. J., Lemaire, R., Yuan, Z., & Keller, J. (2003). Simultaneous nitrification, denitrification, and phosphorus removal in a lab-scale sequencing batch reactor. *Biotechnology Bioengineering, 84*, 170–178.

Zhang, T., Shao, M. F., & Ye, L. (2012). 454 pyrosequencing reveals bacterial diversity of activated sludge from 14 sewage treatment plants. *ISME Journal, 6*, 1137–1147.

Zheng, H., Hanaki, K., & Matsuo, T. (1994). Production of nitrous oxide gas during nitrification of wasterwater. *Water Science and Technology, 30*, 133–141.

Zhou, Y., Pijuan, M., Zeng, R. J., & Yuan, Z. G. (2008). Free nitrous acid inhibition on nitrous oxide reduction by a denitrifying-enhanced biological phosphorus removal sludge. *Environmental Science and Technology, 42*, 8260–8265.

Zhu, X., & Chen, Y. (2011). Reduction of N_2O and NO generation in anaerobic-aerobic (low dissolved oxygen) biological wastewater treatment process by using sludge alkaline fermentation liquid. *Environmental Science and Technology, 45*, 2137–2143.

Chapter 2
Selection and Identification of Two High-Efficient Aerobic Denitrifying Bacteria

2.1 Introduction

In the 1980s, the first aerobic denitrifier *Thiosphaera pantotropha* (now renamed as *Paracoccus denitrificans*) was isolated from a sulfur removing denitrification system, which could reduce nitrate rapidly even at 90% saturation of dissolved oxygen in water (Robertson and Kuenen 1984). This characteristic provides a reliable way to achieve simultaneous nitrification and denitrification in a single reactor, which has attracted wide attention of researchers. Afterwards, more and more aerobic denitrifying bacteria were isolated, such as *Alcaligenes*, *Pseudomonas*, *Achromobacter*, *Comamonas*, and *Bacillus* strains, which have excellent aerobic denitrification performance (Chen and Ni 2011; Chen and Strous 2013). However, due to the sensitivity of N_2O reductase to O_2, most strains produce N_2O rather than N_2 as the final denitrification product, which greatly reduces the application potential of aerobic denitrifying bacteria.

In this chapter, bacterial strains capable of reducing N_2O efficiently during aerobic denitrification were selected from the existing highly effective bacterial strains that were previously isolated in the lab. Then their phylogenetic affiliations were identified by 16S rDNA sequence analysis.

2.2 Materials and Methods

2.2.1 Source of Bacterial Strains

The source of high-efficient bacterial strains in this study was 14 aerobic denitrifying bacteria strains isolated from a steadily operated landfill leachate treatment plants in the previous study in our lab.

M. Zheng, *Nitrogen Removal Characteristics of Aerobic Denitrifying Bacteria and Their Applications in Nitrogen Oxides Emission Mitigation*, Springer Theses,
https://doi.org/10.1007/978-981-13-2432-1_2

2.2.2 Culture Medium

The composition of medium used for denitrification performance evaluation (DM) was as follows (per liter): 8.45 g $C_4H_4Na_2O_4{\cdot}6H_2O$, 0.63 g NH_4Cl, 0.61 g Na $^{15}NO_3$ (99 atom% abundance), 1.76 g $K_2HPO_4{\cdot}3H_2O$, 0.20 g $MgSO_4{\cdot}7H_2O$, 0.02 g $CaCl_2$, 0.005 g $FeSO_4{\cdot}7H_2O$, and 0.1 mL trace elements. $Na^{15}NO_3$ labeled with isotopic ^{15}N was used to avoid the interference of atmospheric N_2 and ensure the precise quantitation of produced $^{15}N_2$.

The composition of medium used for heterotrophic nitrification performance evaluation was as follows (per liter): 5.63 g $C_4H_4Na_2O_4{\cdot}6H_2O$, 0.38 g $^{15}NH_4Cl$, 1.76 g $K_2HPO_4{\cdot}3H_2O$, 0.20 g $MgSO_4{\cdot}7H_2O$, 0.02 g $CaCl_2$, 0.005 g $FeSO_4{\cdot}7H_2O$, and 0.1 mL trace elements. $^{15}NH_4Cl$ labeled with isotopic ^{15}N was applied to ensure the precise quantitation of produced $^{15}N_2$.

The composition of trace elements solution was as follows (per liter): 3.5 g EDTA, 2.0 g $ZnSO_4{\cdot}7H_2O$, 1.0 g $CuSO_4{\cdot}5H_2O$, 2.0 g $MnSO_4{\cdot}7H_2O$, 0.9 g Co $(NO_3)_2{\cdot}6H_2O$, 1.0 g H_3BO_3, and 1.0 g Na_2MoO_4.

All media were adjusted to pH 7.5 and autoclaved for sterility at 121 °C for 30 min before use.

2.2.3 Selection of Aerobic Denitrifying Strains

The 14 aerobic denitrifying bacterial strains were cultured to late exponential growth phase, and 10 mL of bacterial suspension was collected, washed, and resuspended in DD H_2O and then inoculated into 50 mL DM in 300 mL glass serum bottles with gas-impermeable rubber stoppers. The bottles were incubated in rotary shaker at 150 rpm and 30 °C. 5 mL aliquots were withdrawn periodically from the bottles and centrifuged. The supernatant was stored in a refrigerator at 4 °C and obtained for NH_4^+–N, NO_2^-–N, NO_3^-–N analysis. At the same time, 100 μL gas samples were collected from the headspaces of the bottles for the determination of N_2O and N_2 concentrations. Strains that performed good aerobic denitrification capability with low N_2O accumulation were selected.

2.2.4 Identification of Aerobic Denitrifying Strains

The morphology of selected strains was observed using an environmental scanning electron microscope (Quanta 200FEG, FEI, USA). Total bacterial DNA was extracted from 1 mL bacterial suspension by genomic DNA extraction kit, and the 16S rDNA was amplified by universal primers F27 and R1492. PCR protocol were initial denaturation at 95 °C for 10 min, 40 cycles of denaturation at 95 °C for 10 s, annealing at 55 °C for 20 s, and extension at 72 °C for 1 min. The PCR products were separated by agarose gel electrophoresis, where the size of the fragments was checked by the agarose gel electrophoresis. The PCR products were sent to Invitrogen Biotechnology Co., Ltd. for sequencing. The obtained sequences were analyzed and aligned through Basic Local Alignment Search Tool (BLAST) in NCBI database. Finally, a neighbor-joining phylogenetic tree was constructed based on the 16S rRNA gene sequences of the isolated strains and the reference sequences in MEGA 5.05 software with 1000 bootstrap replicates and the maximum composite likelihood model.

2.2.5 Analytical Methods

All reagents used in this study were analytical grade, unless otherwise specified. The bacterial growth was determined by a spectrophotometer (UV-1800, Shimadzu, Japan) at 600 nm (OD600). Dissolved oxygen (DO) was measured using a DO analyzer (550A, YSI, USA). NH_4^+–N, NO_2^-–N, and NO_3^-–N were measured using a discrete autoanalyzer (Smartchem 200, AMS, Italy). O_2 and N_2O were analyzed by a gas chromatography (GC) equipped with an electron capture detector (ECD). $^{15}N_2$ was determined by gas chromatography-mass spectrometer (GC/MS) with the charge-mass ratio of 30. Standard curves (Fig. S1) were prepared with pure O_2, N_2O, and $^{15}N_2$ (99 atom% abundance) diluted to different concentrations.

2.3 Results and Discussion

2.3.1 Aerobic Denitrification Performance of the Strains

The 14 aerobic denitrifying bacterial strains were cultured in DM for evaluation of aerobic denitrification performance. Bacterial growth and NO_3^-–N and TN removal rates after 12 h cultivation were shown in Table 2.1. As can be seen from the data, the 14 strains could grow rapidly in DM since the OD_{600} reached 0.91–1.73 after 12 h. However, the removal rate of NO_3^-–N and TN was significantly different; strains HD2–1, BD1-1, BD1-2, BD5, and BD6 were not able to conduct good aerobic denitrification with TN removal rates below 10%. At the same time, the other nine strains could remove more than 80% of TN. Since respiring oxygen and nitrate simultaneously is one of the most significant characteristics of aerobic denitrification bacteria, this result shows that these nine strains have good aerobic denitrification performance.

The complete denitrification process is the process of NO_3^-–N being reduced to N_2 via NO_2^-–N, NO, and N_2O, and the degradation of nitrate is the first step of denitrification process. In order to assess the denitrification capacity, it is particularly important to determine the gaseous intermediate and final product. As can been seen from the serum bottle experiments, the denitrification products of strain HD4-1 were

Table 2.1 Aerobic denitrification and gaseous products of the 14 aerobic denitrifying strains

	Shake flask experiment				Sealed bottle experiment			
Strains	Initial NO_3-N	Maximal OD_{600}	NO_3-N removal (%)	TN removal (%)	Residual O_2 (%)	Maximal N_2O (mg L^{-1})	12 h N_2O (mg L^{-1})	12 h N_2 (mg L^{-1})
HD1-1	99.46	1.55	100.00	89.99	7.76	85.37	84.92	0.13
HD1-2	96.83	1.60	99.02	99.00	5.60	95.58	92.85	0.13
HD2-1	102.05	1.27	5.57	5.43	4.96	0.02	0.01	0.12
HD2-2	95.65	1.58	100.00	82.42	4.26	87.59	87.59	0.11
HD3-1	95.91	1.57	99.08	99.01	4.32	85.65	85.65	0.12
HD3-2	94.92	1.58	99.77	87.37	4.64	83.70	74.79	0.12
HD4-1	102.63	0.91	100.00	87.73	7.20	0.11	0.04	95.54
HD4-2	103.65	1.73	100.00	100.00	4.80	66.20	52.31	44.11
BD1-1	98.06	1.08	4.83	4.67	5.65	0.05	0.02	0.11
BD1-2	96.20	1.13	16.17	2.99	8.66	0.05	0.02	0.09
BD3	97.68	1.53	100.00	81.39	5.78	85.44	78.12	0.13
BD4	98.95	1.57	98.31	98.31	6.04	87.50	81.64	0.13
BD5	102.05	1.14	71.43	6.87	8.99	0.35	0.28	0.09
BD6	99.98	1.27	1.10	0.73	3.71	0.04	0.03	0.12

Note: The initial concentration of NO_3^-–N in the shake flask experiment was set at 100 mg L^{-1}; the initial concentration of NO_3^-–N in the sealed bottle experiment was set at 80 mg L^{-1}

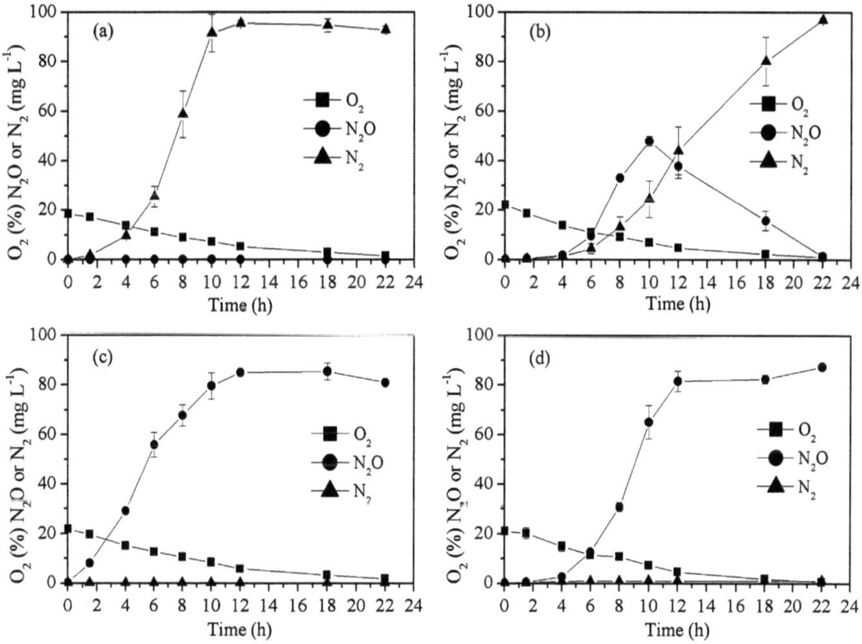

Fig. 2.1 Variations of O_2, N_2O, and N_2 during aerobic denitrification by strain HD4-1 (**a**), HD4-2 (**b**), HD1-1 (**c**), and PAO1 (**d**). (Reprinted from Zheng et al. (2014), with permission from Elsevier)

almost only N_2, and the denitrification products of HD4-2 were N_2O and N_2, while the denitrification products of the other seven strains were mainly N_2O. As for the nitrogen balance problem in some experiments, it may be due to the error caused by different nitrogen detection methods and the N_2O concentration that did not reach equilibrium between gas and liquid.

The denitrification process of strains HD4-1, HD4-2, and HD1-1 was shown in Fig. 2.1. As can been seen, there was no obvious accumulation of N_2O during the whole process (the highest accumulation was 0.08 mg L^{-1}) in strain HD4-1, while N_2 increased gradually and reached its maximum value at 10 h. In strain HD4-2, N_2O and N_2 increased significantly during the first 10 h of denitrification, and after 10 h N_2O began to decrease gradually and finally completely reduced to N_2 at 22 h. Strain HD1-1 was similar to most of the other strains with denitrifying ability, N_2O gradually increased during the whole denitrification process, and there was no obvious accumulation of N_2. The denitrification performance of the representative denitrifying strain PAO1 was also studied. The results showed that the final denitrification product was still N_2O (N_2 accumulation was 1.08 mg L^{-1}), which proved that the inhibitory effect of O_2 on N_2O reductase was the main reason to prevent N_2O reduced to N_2.

Therefore, considering the unique properties of strain HD4-1 and HD4-2 reducing N_2O to N_2 under aerobic conditions, these two strains were further studied in detail and renamed as PCN-1 and PCN-2.

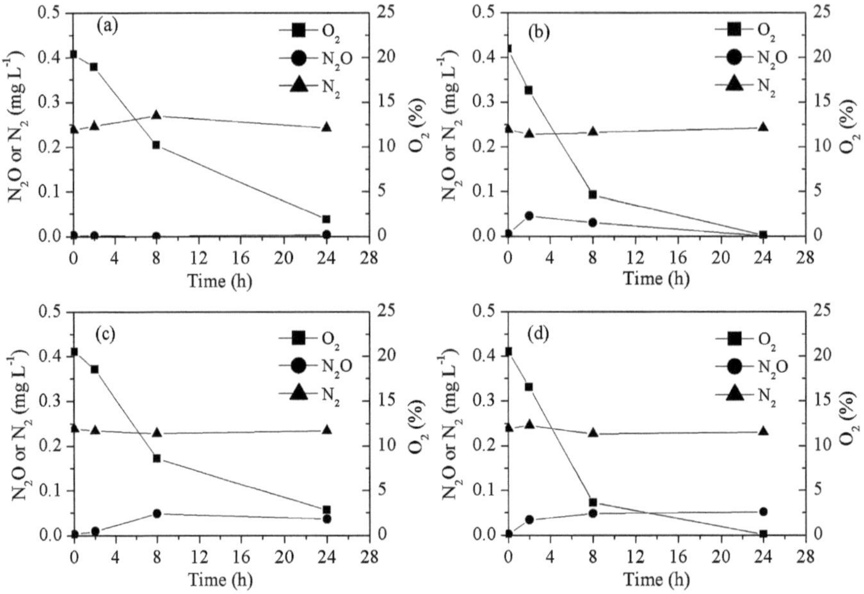

Fig. 2.2 Variations of O_2, N_2O, and N_2 during heterotrophic nitrification with ammonia as sole nitrogen source by strain HD4-1 (**a**), HD4-2 (**b**), HD1-1 (**c**), and PAO1 (**d**). (Reprinted from Zheng et al. (2014), with permission from Elsevier)

2.3.2 Heterotrophic Nitrification Performance of the Two Strains

It is noteworthy that some aerobic denitrifying bacteria also have the function of heterotrophic nitrification, which means that the produced NO, N_2O, and N_2 also could come from the heterotrophic nitrification-aerobic denitrification of ammonia nitrogen (Otte et al. 1996; Zhang et al. 2012). As can be seen from Fig. 2.2, NO, N_2O, and N_2 accumulation was not detected in the whole culture process when the four strains were supplied with $^{15}NH_4Cl$ as the sole nitrogen source. Therefore, it can be deduced that the four strains have no function of heterotrophic nitrification, and all the three gases accumulated during aerobic denitrification process come from the reduction process of nitrate.

2.3.3 Species Identification of Strains PCN-1 and PCN-2

As observed by physiological experiments and scanning electron microscopy (SEM), PCN-1 and PCN-2 were all Gram-negative bacteria and short rod-shaped,

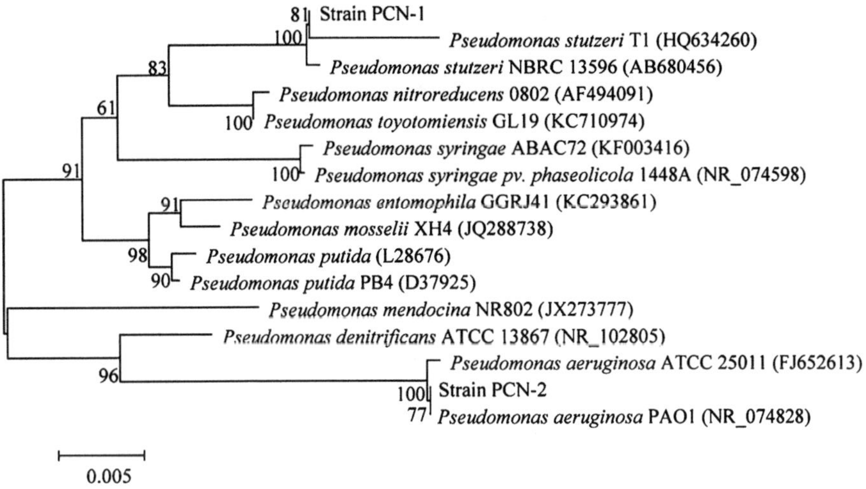

Fig. 2.3 Phylogenetic tree based on 16S rRNA gene sequence of strain PCN-1, strain PCN-2, and other reference sequences

free of spore, movable and monopolar flagellum. The sizes of PCN-1 and PCN-2 were (0.5–0.7) μm × (1.3–1.9) μm and (0.7–1.0) μm × (2.3–2.8) μm, respectively.

The 16S rRNA of two strains was amplified by polymerase chain reaction (PCR) and1404 and 1456 base pairs were obtained, with accession number of KF589921 and KF589922 in the National Center for Biotechnology Information (NCBI) GenBank database. The phylogenetic tree analysis shows that PCN-1 and *P. stutzeri* T1 were found to be 99% similar and PCN-2 and *P. aeruginosa* PAO1 reached 99% similarity. Therefore, strains PCN-1 and PCN-2 were identified as *Pseudomonas stutzeri* PCN-1 and *Pseudomonas aeruginosa* PCN-2, respectively (Fig. 2.3).

2.4 Summary

In this chapter, the following conclusions were obtained:

1. Denitrification capacity of 14 aerobic denitrifying bacteria was compared, and two strains capable of reducing N_2O to N_2 were selected and named as PCN-1 and PCN-2. Further study showed that all the gaseous products were derived from aerobic denitrification process.
2. Strain PCN-1 and PCN-2 was identified as *Pseudomonas stutzeri* and *Pseudomonas aeruginosa* by 16S rRNA phylogenetic tree analysis.

References

Chen, Q., & Ni, J. R. (2011). Heterotrophic nitrification-aerobic denitrification by novel isolated bacteria. *Journal of Industrial Microbiology and Biotechnology, 38*, 1305–1310.

Chen, J. W., & Strous, M. (2013). Denitrification and aerobic respiration, hybrid electron transport chains and co-evolution. *Biochimica et Biophysica Acta, 1827*, 136–144.

Otte, S., Grobben, N. G., Robertson, L. A., Jetten, M. S., & Kuenen, J. G. (1996). Nitrous oxide production by *Alcaligenes faecalis* under transient and dynamic aerobic and anaerobic conditions. *Applied and Environmental Microbiology, 62*, 2421–2426.

Robertson, L. A., & Kuenen, J. G. (1984). Aerobic denitrification: A controversy revived. *Archives of Microbiology, 139*, 351–354.

Zhang, Q. L., Liu, Y., Ai, G. M., Miao, L. L., Zheng, H. Y., & Liu, Z. P. (2012). The characteristics of a novel heterotrophic nitrification-aerobic denitrification bacterium, *Bacillus methylotrophicus* strain L7. *Bioresource Technology, 108*, 35–44.

Zheng, M. S., He, D., Ma, T., Chen, Q., Liu, S. T., Ahmad, M., Gui, M. Y., & Ni, J. R. (2014). Reducing NO and N_2O emission during aerobic denitrification by newly isolated Pseudomonas stutzeri PCN-1. *Bioresource Technology, 162*, 80–88.

Chapter 3
Aerobic Denitrification Characteristics and Mechanism of *Pseudomonas stutzeri* PCN-1

3.1 Introduction

More attention has been paid on the phenomenon of NO and N_2O accumulation in the process of denitrification by the researchers. For example, the typical aerobic denitrifying bacterium *P. denitrificans* LMD 92.63 produced only N_2O under the condition of a saturated dissolved oxygen concentration of 95% (Arts et al. 1995); strain *P. denitrificans* JCM 20620 released N_2O as the final denitrification product as long as O_2 is present (Miyahara et al. 2010); N_2O had been produced by the strain *Alcaligenes faecalis* TUD, which accounted for 5% of the total nitrogen removal under the condition of 5% saturation dissolved oxygen concentration (Otte et al. 1996). In recent years, a number of aerobic denitrifying bacteria have been successfully separated, but the quantitative analysis of accumulation of NO and N_2O has rarely been reported (Wan et al. 2011; Yang et al. 2011; Yao et al. 2013).

After confirming the regularity of N_2O accumulation and utilization, the quantitative analysis of the denitrification function gene revealed the mechanism of its unique denitrification. Influence factors such as O_2 concentration, types of carbon source, pH, and C/N ratio on nitrification, and the accumulation of NO and N_2O were also studied in this research. Finally, its strengthened capacity of nitrogen and nitrogen oxide emissions reduction on sewage treatment was confirmed through bioaugmentation shake flask experiments.

M. Zheng, *Nitrogen Removal Characteristics of Aerobic Denitrifying Bacteria and Their Applications in Nitrogen Oxides Emission Mitigation*, Springer Theses,
https://doi.org/10.1007/978-981-13-2432-1_3

3.2 Materials and Methods

3.2.1 *Medium*

The denitrification test medium formulation (per liter) consisted of 8.45 g $C_4H_4Na_2O_4{\cdot}6H_2O$, 0.63 g NH_4Cl, 0.61 g $NaNO_3$ or 0.39 g $NaNO_2$, 1.76 g $K_2HPO_4{\cdot}3H_2O$, 0.20 g $MgSO_4{\cdot}7H_2O$, 0.02 g $CaCl_2$, 0.005 g $FeSO_4{\cdot}7H_2O$, and 0.1 ml trace elements solution.

Media formulation used in activated sludge shock test (per liter) consisted of 1.71 g CH_3COONa, 0.19 g NH_4Cl, 0.44 g $K_2HPO_4{\cdot}3H_2O$, 0.20 g $MgSO_4{\cdot}7H_2O$, 0.02 g $CaCl_2$, 0.005 g $FeSO_4{\cdot}7H_2O$, and 0.1 ml trace elements solution. When 0.39 g L^{-1} $NaNO_2$ or 0.61 g L^{-1} $NaNO_3$ was added to the simulated wastewater, the amount of CH_3COONa was increased to 4.45 or 5.13 g L^{-1}.

3.2.2 *Growth Measurement of Strain PCN-1*

The strain PCN-1 was inoculated into LB medium undergoing 24 h incubation at the setting of 30 °C and 150 fpm. A certain amount of culture liquid was then taken and diluted into bacteria suspension of different concentrations for their corresponding optical density OD_{600} measurement. 9 mL bacterial suspension of each concentration was added to the priorly dried and weighed centrifuge tubes. Centrifugation was carried out at 8000 rpm for 5 min, and the supernatant was then discarded. The cells were dried at 105 °C until constant weight. Dispersion method was applied to calculate the dry weight of bacteria, and the standard curve of OD_{600} was drawn (Fig. S2).

3.2.3 *Study on Aerobic Denitrification Performance of Strain PCN-1*

The denitrification characteristics of the strain PCN-1 were investigated in a 300 mL closed anaerobic flask with the same methods in Sect. 2.2.3. When N_2O was used as denitrification substrate, 1.2 mL of pure N_2O was injected into the top of the anaerobic bottle before inoculation.

In the study of the effect of oxygen concentration on the denitrification and N_2O production, the top space of the anaerobic bottle was replaced by high purity helium, and O_2 was substituted for different volumes (0%, 10%, 20%, 50%, and 100%) of helium. Experiments were carried out simultaneously in jars to determine the dissolved oxygen concentration in the culture medium. The initial concentration of ammonia and nitrogen increased from 150 mg L^{-1} to 300 mg L^{-1} to meet the requirement of cell assimilation. For the inspection of the carbon source, sodium succinate, sodium citrate, sodium acetate, glucose, sucrose, and so on were selected on the conditions initial of C/N = 10 and pH = 8. In the experiment of pH and C/N, sodium acetate was used as carbon source, and N_2O production was investigated under a variable pH value of 6, 7, 8, 9, and 10 with C/N equal to 10 and a variable C/N of 2, 4, 6, 10, and 15 with a pH value equal to 8.

In order to study the application of the strain in the wastewater treatment, 7.5 and 15 mL cell culture medium (OD_{600} = 1.053) were collected and centrifuged and resuspended in a 300 mL anaerobic bottle containing 50 mL activated sludge (MLVSS = 1208 mg L^{-1}), and the corresponding bacterial concentrations were 126 mg VSS L^{-1} and 252 mg VSS L^{-1}, respectively. The corresponding inoculum size was 15% and 30% (v/v%), respectively. At the same time, in the contrast group, a certain amount of activated sludge was added without inoculation, so that the total amount of bacteria and the experimental group maintains consistency.

In addition to examine the effect of O_2 concentration on the experiment, the rest of the top air in anaerobic test bottle is no longer replaced (O_2 concentration of about 21%). The anaerobe was inoculated, shaked, and cultured under the condition of 150 rpm and 30 °C. N_2O, O_2, and N_2 were measured at a constant time by using a valve syringe, and 2 mL of gas was injected into 2 L of helium-filled airbags. At the same time, 1 mL of bacterial suspension was sampled with a syringe and centrifuged at 8000 rpm for 4 min at 4 °C. The supernatant was used to analyze the concentrations of ammonia nitrogen, nitrite, and nitrate. In the denitrification experiments with nitric nitrogen as substrate, the cells after centrifugation were stored at −80 °C for RNA extraction.

3.2.4 Real-Time Quantitative PCR Analysis of Denitrification Genes

The expression of denitrifying gene was quantitatively analyzed by a two-step method of reverse transcription and real-time PCR. Total RNA extraction and cDNA synthesis were performed using Total RNA Extraction Kit and Rapid Reverse Transcription Kit (Tiangen, China). Quantitative PCR was performed using the SYBR mixed dye (Applied Biosystems, USA) and the corresponding denitrifying gene nirS, cnorB, and nosZ primers (Table 3.1). The V3 region of 16S rRNA was used as internal reference to unify the difference in the total amount of cDNA in each PCR tube. The standard curve was prepared by standard dilution of the prepared standard plasmid and quantitatively analyzed simultaneously with the sample. The final reaction was performed in a quantitative PCR system (Applied Biosystems 7300, USA) under the following conditions: 95 °C pre-denaturation 10 min, 40 cycles of 95 °C denaturation 10 s, and 60 °C (nirS and nosZ) or 56 °C (cnorB) annealing extended 1 min.

Table 3.1 PCR primers used for 16S rRNA gene and denitrifying genes amplification

Gene name	Primer name	Primer sequence(5′-3′)	Length (bp)	References
16S rRNA	F27	AGAGTTTGATCMTGGCTCAG	1465	Yao et al. (2013)
	R1492	TTGGYTACCTTGTTACGACT		
16S V3	F341	CCTACGGGAGGCAGCAG	177	Muyzer et al. (1993)
	R518	ATTACCGCGGCTGCTGG		
nirS	nirS cd3aF	GTSAACGTSAAGGARACSGG	425	Throback et al. (2004)
	nirS R3cd	GASTTCGGRTGSGTCTTGA		
cnorB*	cnorB Z1F	CGTCGGTCAGATCCTCTTCG	657	
	cnorB Z1R	GCGATGATCACGTAGAGCCA		
nosZ	nosZ 1527F	CGCTGTTCHTCGACAGYCA	246	Throback et al. (2004)
	nosZ 1773R	ATRTCGATCARCTGBTCGTT		

*The sequence of cnorB gene of P. stutzeri A1501 was designed as a template in the NCBI database

3.2.5 *Analytical Methods*

The optical density (OD_{600}), NH_4^+-N, NO_2^--N and NO_3^--N, O_2, N_2O, and N_2 concentrations were determined in the same way as 2.2.5.

Dissolved oxygen (DO) was measured using a DO analyzer (550A, YSI, USA). The NO analysis was performed by injecting 2 mL of the collected gas sample into a 2 L air-filled bag, which contained helium, and measured it using an NO-N_2O-NO_x analyzer (Model 42i, Thermo Fisher Scientific Inc., USA).

The concentration (ppm) of NO, N_2O, and N_2 in the gaseous products was first obtained by direct reading or standard curve and then converted to mass concentration, (under the condition of 30 °C and 1 bar, 1 ppm NO $= 0.56\ \mu g\ L^{-1}$, 1 ppm $N_2O = 1.126\ \mu g\ L^{-1}$). In order to observe the changes of denitrification products and substrate concentration more clearly, the unified volume of liquid was represented as the standard. O_2 was expressed in percentage (%).

ANOVA and Tukey test of SPSS 13.0 software were applied for significance analysis, where significant difference was determined when $P < 0.05$.

3.3 Results and Discussion

3.3.1 *Aerobic Denitrification Performance of Strain PCN-1 to Nitrate*

The nitrogen transformation process and the process of growth and oxygen consumption during the aerobic denitrification of strain PCN-1 are shown in Fig. 3.1.

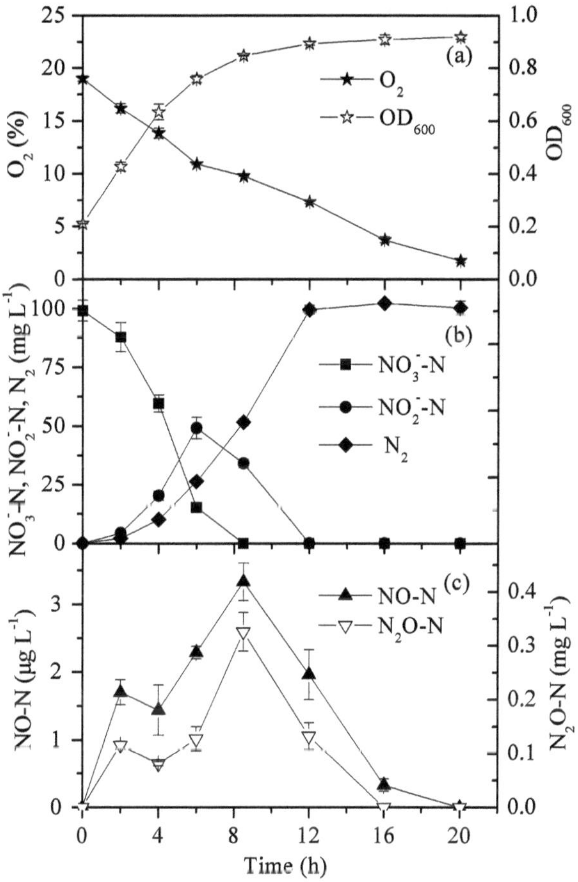

Fig. 3.1 Aerobic denitrification of strain PCN-1 with nitrate as substrate. (Reprinted from Zheng et al. (2014), with permission from Elsevier)

After the inoculation, the strain began to grow quickly and reached the stationary phase at 12 h. The initial concentration of 100 mg L^{-1} was also used by the bacteria after inoculation and decreased rapidly in the logarithmic phase of bacterial growth. The average removal rate of nitrate nitrogen that can be calculated was 11.66 mg L^{-1} h^{-1}, which was faster than that of most aerobic denitrifying bacteria reported in the past (Table 1.1). It can be seen that O_2 was still 9.77% of the remaining time, which indicated that the strain PCN-1 could conduct aerobic respiration and nitrate respiration at the same time, and it played an important role in aerobic denitrification.

Intermediate (NO, N_2O) and final product (N_2) of the denitrification process were detected at the first sampling point of 2 h, which indicated that the denitrifying enzymes in the four stages of denitrification have been synthesized and exerted after inoculation within 2 h. Accumulation of nitrite nitrogen increased rapidly with the decrease of nitrate nitrogen and reached a maximum of 49.10 mg L^{-1} in 6 h, but nitrate nitrogen began to decline rapidly after it had been completely consumed and fully utilized in 12 h. NO and N_2O showed the same trend, with a relatively high accumulation in 2 h, probably due to the pressure of survival of the new culture after inoculation; a similar phenomenon has been reported in the past (Kester et al. 1997). After a fluctuation, NO and N_2O reached the maximum values of 3.33 μg L^{-1} and 0.33 mg L^{-1} in 8.5 h but only 0.003% and 0.33% of the removal amount of nitrate nitrogen, respectively. Both NO and N_2O were also quickly used after nitrate nitrogen and nitrite nitrogen were completely consumed. Finally, N_2 was maintained at a concentration level of about 100 mg L^{-1}, implying that the initial nitrogen was ultimately converted to N_2. Although the ammonia nitrogen in the culture medium was also used by the strain, the denitrification gas was all derived from the nitrification process, because the strain had no heterotrophic nitrification.

This result implies that not only the denitrification process of nitrate and nitrite was not inhibited by O_2, but also the reduction of NO and N_2O was not inhibited by O_2, which is in sharp contrast to the conventional aerobic denitrifying bacteria. For example, the expression of the cnorB gene was not detected in strain *P. mandelii* under the conditions of aerobic denitrifying bacteria (Saleh-Lakha et al. 2008); strains *A. eutrophus* LMD 82.41 and *P. stutzeri* LMAU P12 were stable at less than 1% saturated dissolved oxygen, but the accumulation of NO reached 0.39% and 0.13%, respectively, which was much higher than 0.003% of PCN-1 produced in this study. The denitrifying strain *P. denitrificans* LMD 92.63 produced 95.79% of the N_2O removal under the condition of 20% saturation dissolved oxygen, which reached 100% when dissolved oxygen was 95% saturated (Arts et al. 1995). *P. denitrificans* strain JCM 20620 released as the final denitrification product in the presence of O_2, implying that N_2O reduction was also significantly inhibited by the presence of trace O_2 (Miyahara et al. 2010). In recent years, more aerobic denitrifiers have been successfully isolated, but few studies investigated the accumulation of NO and N_2O (Wan et al. 2011; Yang et al. 2011; Yao et al. 2013). Although there is no quantitative analysis, significant N_2O accumulation was observed (Zhang et al. 2012a, b). Therefore, it can be concluded that the excellent aerobic denitrification and non-accumulation of NO and N_2O of strain PCN-1 provided a reliable way for enhanced denitrification and greenhouse gas emission reduction during wastewater treatment.

3.3.2 Aerobic Denitrification Performance of Strain PCN-1 with Nitrite and N_2O

When nitrite and N_2O were used as denitrifying substrates, the aerobic denitrification characteristics of strain PCN-1 were shown in Fig. 3.2. It was the same as the experiment which used nitrate as the substrate, where 150 mg L^{-1} ammonia nitrogen was added into the medium to provide nitrogen source for bacterial assimilation growth. It can be seen that the rate of nitrite removal was 12.80 mg L^{-1} h^{-1} when the strain was rapidly grown and the nitrite was quickly consumed and removed completely in 7 h after the start of culture (Fig. 3.2a), which was much higher than previous report, 0.76 mg L^{-1} h^{-1} reached by aerobic denitrifier *Pseudomonas sp.*

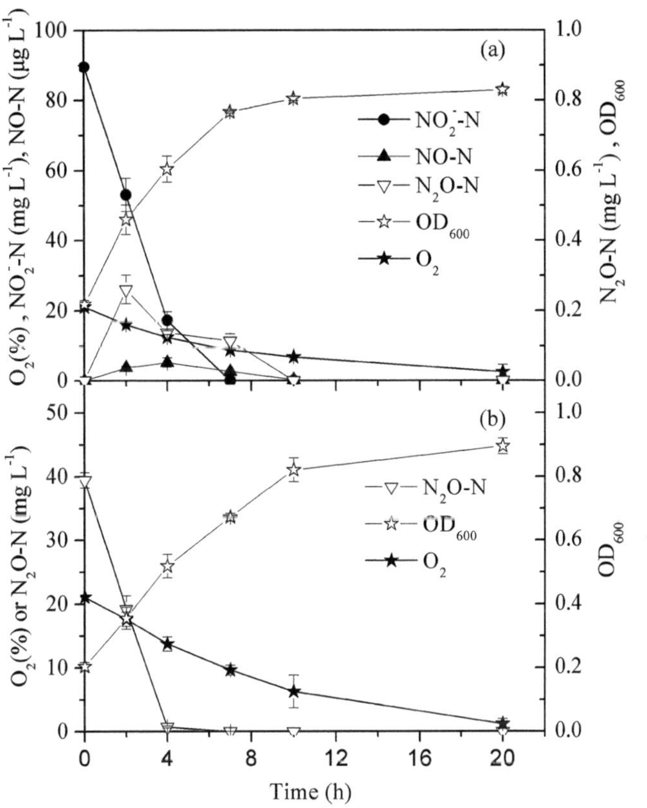

Fig. 3.2 Aerobic denitrification performances of strain PCN-1 with nitrite (**a**) or N_2O (**b**) as substrate. (Reprinted from Zheng et al. (2014), with permission from Elsevier)

Yy7 and 0.24 mg L^{-1} h^{-1} reached by *Bacillus methylotrophicus* L7 (Wan et al. 2011; Zhang et al. 2012a, b). According to previous studies, high concentrations of nitrite had inhibitory effects on bacterial growth and denitrification (Bergaust et al. 2008; Wan et al. 2011). However, the strain PCN-1 showed good growth and denitrification characteristics, which indicated that it could tolerate higher concentrations of nitrite. It could be seen that the maximum accumulation of NO and N_2O is 5.19 μg L^{-1} and 0.26 mg L^{-1} in the process of denitrification, only accounting for 0.006% and 0.29% of the removal of nitrite, which was equivalent with the accumulation amount of the substrate when denitrifying. In contrast, aerobic denitrifier *Pseudomonas sp.* Yy7 has significant inhibitory effects at the initial nitrite concentration of 50 mg L^{-1}, and the final nitrite removal rate was only 46% (Wan et al. 2011). When the strain *agrobacterium tumefaciens* C58 was at the initial concentration of 2 mM of nitrite and 7% of O_2, the inhibition of nitrite to NO reductase resulted in a NO accumulation of up to 40 μM, accounting for 78.26% of the nitrite removal (Bergaust et al. 2008).

When N_2O was used as the denitrifying substrate, the removal rate of N_2O-N was 98.13%, and the average removal rate of nitrogen was 9.67 mg L^{-1} h^{-1} for the initial 39.42 mg L^{-1} in 4 h (Fig. 3.2b). It can be seen that this rate was comparable to the rate of nitrite or nitrate as substrates, which reflected the fact that N_2O reduction could not be inhibited by O_2 as with nitrate or nitrite reduction, and it explained the phenomenon that N_2O accumulation is so low during denitrification. The copper core (CuZ) in the N_2O reductase was easily inhibited by O_2, leading to an interruption of the electron transport chain to N_2O, which was the mechanism that N_2O reduction is inhibited by O_2. For example, when the culture conditions were changed from anaerobic to 25% saturated with dissolved oxygen, the reduction effect of aerobic denitrifying bacterium *A. faecalis* TUD to N_2O was immediately terminated, which resulted in significant N_2O emissions (Otte et al. 1996). Denitrificans JCM 20620 could not denitrify with N_2O as substrate in the presence of O_2. In this study, strain PCN-1 showed a strong capacity of aerobic N_2O reduction, which further proved the great potential of the strain to N_2O emission in the biological nitrogen removal system.

3.3.3 Expression of Denitrifying Functional Genes in Strain PCN-1

The expression levels of denitrifying genes nirS, cnorB, and nosZ were analyzed by reverse transcription-fluorescence quantitative real-time PCR, and the strain PCN-1 use nitrite as substrate in this process. The results are shown in Fig. 3.3. It can be seen that all three genes were successfully expressed during aerobic denitrification,

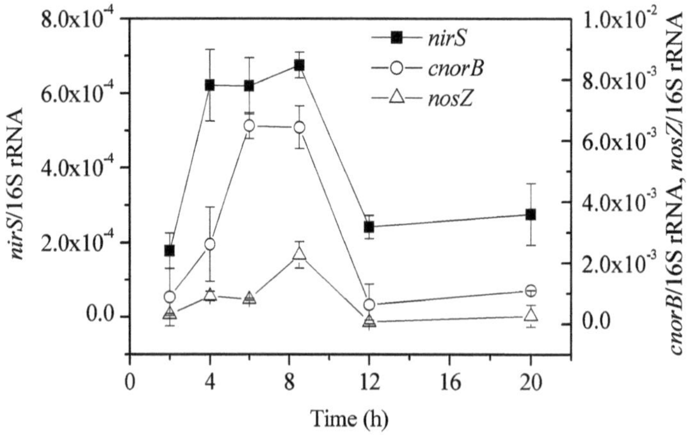

Fig. 3.3 Expression of *nirS*, *cnorB*, and *nosZ* relative to 16S rRNA when nitrate was taken as denitrifying substrate of strain PCN-1. (Reprinted from Zheng et al. (2014), with permission from Elsevier)

indicating that they were involved in the denitrification process. The expression of denitrifying gene of *P. fluorescens* C7R12 and the cnorB gene of strain *P. mandelii* could not be expressed under aerobic condition. The strain PCN-1 showed a unique aerobic denitrification function.

The expression of nirS increased rapidly after inoculation; it reached the maximum in 4–8.5 h, which corresponded to the accumulation of nitrite. Since the expression of the nirS gene required the presence of nitrate or nitrite (Hartig and Zumft 1999), the expression of nirS rapidly decreased in 12 h. The expression of cnorB gene reached the highest in 6–8.5 h, which corresponded to a relatively high accumulation of NO. This was in agreement with previous studies that NO regulates the expression of denitrifying genes, especially nor genes, to keep NO accumulation at a low level to prevent cells damaging themselves (Spiro 2012). It could also be seen from Fig. 3.3 that the expression of cnorB was about one order of magnitude higher than that of nirS, thus ensuring sufficient NO reductase synthesis to maintain a low NO concentration. As the last process of catalytic denitrification, the expression of nosZ increased gradually and reached the highest level in 8.5 h, which was consistent with the highest rate of N_2 production at the same time.

Interestingly, the expression of three denitrifying genes, nirS, cnorB, and nosZ, showed the same trend; the expressions of all were detected in 2 h and then increased with the accumulation of intermediates. Finally it decreased in 12 h to a lower concentration. It was speculated that the synergistic expression of these three functional genes is the root cause of the low accumulation of NO and N_2O, which was also consistent with the reported nir, nor, and nos genes of *Pseudomonas stutzeri* that form a cluster of genes in DNA (Zumft 1997). The rapid decline of the three genes in 12 h was due to no induction of the absence of substrate and the instability of mRNA; the half-life of Pseudomonas strychni's mRNA was 13 min (Hartig and Zumft 1999).

3.3.4 Denitrification Performances of Strain PCN-1 Under Different Oxygen Conditions

Oxygen concentration played a decisive role in the aerobic denitrification process. Therefore, denitrification and NO and N_2O production performances of strain PCN-1 at different O_2 concentrations were investigated. The results were shown in Fig. 3.4. In the anaerobic bottles, the proportion of O_2 concentrations in the headspace was 0%, 10%, 20%, 50%, and 100%, respectively (Fig. 3.4a). The DO concentrations in the initial medium were 0.03, 3.26, 4.18, 8.29, and 18.39 mg L^{-1}, respectively. Under the initial condition of 0–50% O_2 concentrations, the degradation process of nitrogen had no lag phase and could be degraded completely within 12 h. Although nitric nitrogen was delayed for 6 h under the condition of 100% of O_2 concentrations, the average nitrate removal rate was 5.45 mg L^{-1} h^{-1}, but finally the nitrate could be fully utilized within 16 h. The denitrification rate of

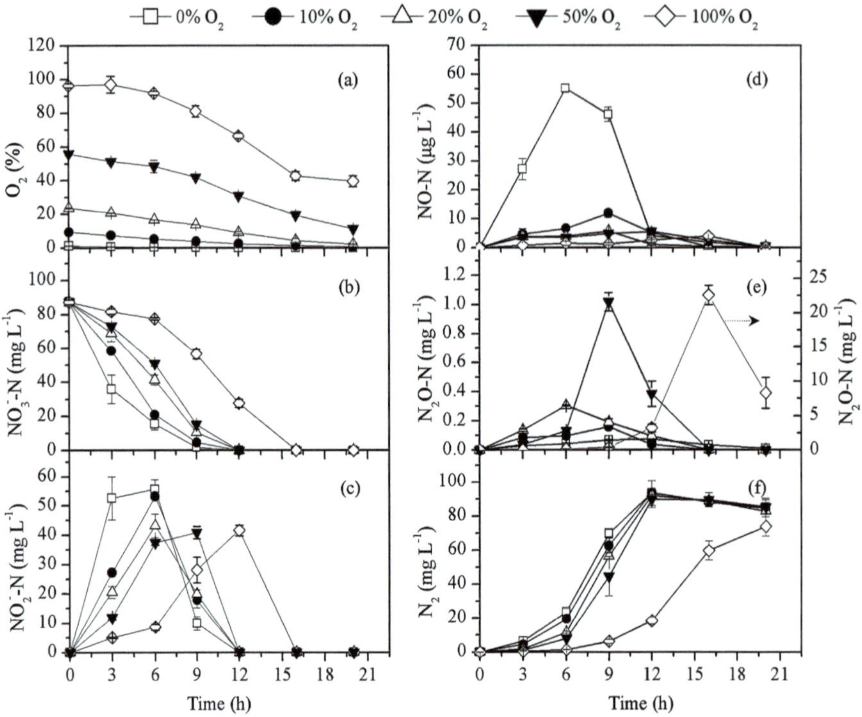

Fig. 3.4 Influence of O_2 concentrations on aerobic denitrification characteristics of strain PCN-1. (Reprinted from Zheng et al. (2014), with permission from Elsevier)

aerobic denitrifying bacteria was much faster than most of that reported under the condition of saturation O_2 concentration (21%) (Chen et al. 2012; Guo et al. 2013), showing good tolerance to high O_2 concentration.

NO accumulation was significantly decreased with increasing O_2 concentration (Fig. 3.4d), and this meant that the activity of NO reductase was not inhibited by higher O_2 concentration. On the contrary, the activity of NO reductase had become much higher. All copper oxygen reductases are believed to reduce NO, indicating that O_2 and NO could be reduced simultaneously under conditions of aerobic (Chen and Strous 2013). Moreover, a globular protein called erythrocyte hemoglobin, which was widely present, could readily convert NO to nitrate under the conditions of aerobic, thereby preventing the cells from NO (Poole and Hughes 2000). Therefore, under the conditions of aerobic, bacteria could transform NO through a variety of ways, which has become one of the inherent advantages of aerobic denitrifying bacteria. The accumulation of N_2O increased with the increasing of O_2 concentration (Fig. 3.4e). However, under the condition of 50% O_2 concentration, the N_2O accumulation was only 1.12% of that of nitrate removal. In the environment of 100% O_2 concentration, the N_2O accumulation reached 22.58 mg L^{-1}, which was equivalent to 25.90% of the removal of nitrate, mainly due to the inhibition of N_2O

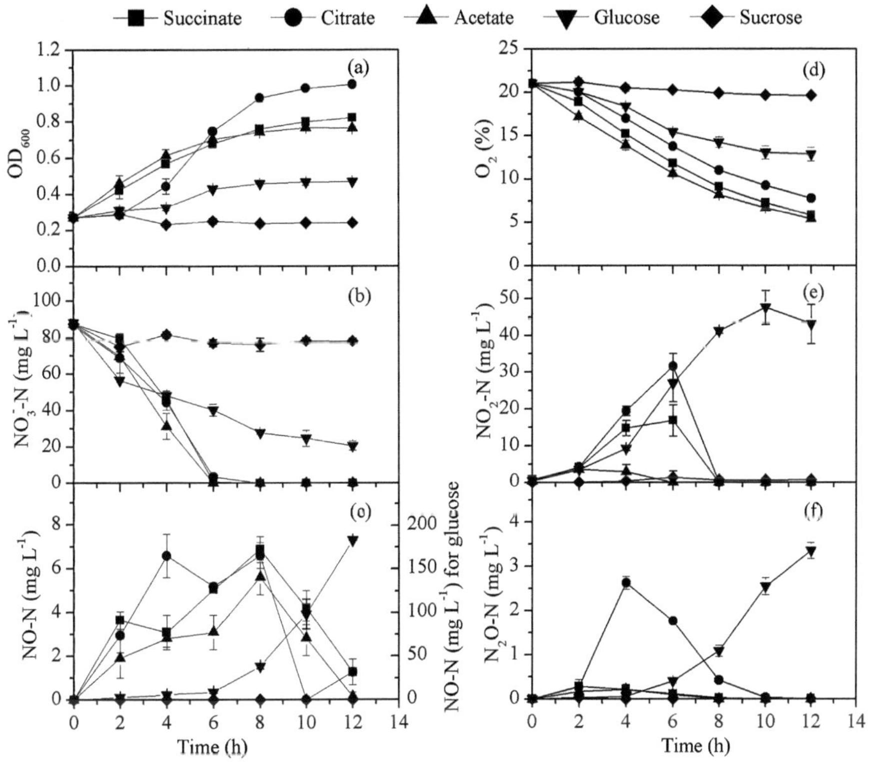

Fig. 3.5 Aerobic denitrification characteristics of strain PCN-1 under different carbon sources. (Reprinted from Zheng et al. (2014), with permission from Elsevier)

enzyme under this extreme condition of O_2. Another reason might be that reactive oxygen species (ROS) might be produced by bacteria in the presence of aerobic respiration, which resulted in the damage to the DNA, RNA, and proteins of the cells (Cabiscol et al. 2000).

3.3.5 Denitrification Performance of Strain PCN-1 Under Different Conditions of Culture

It can be seen from Fig. 3.5 and Table 3.2 that the carbon source had a significant effect on the growth of strain PCN-1 ($P < 0.05$). The strain PCN-1 could grow well under the conditions of sodium succinate, sodium citrate, and sodium acetate as carbon sources and reached 100% nitrification rate in 12 h, but sodium citrate significantly produced much more N_2O. When glucose was used as a carbon source, the growth of bacterial and the removal of nitrite were both not ideal with a large amount of NO, and N_2O was produced, which is likely to come from the

Table 3.2 Aerobic nitrate removal and maximal NO-N and N*O-N production during 12 h incubation under different carbon sources, pH values, and C/N ratios*

Influencing factors		MaximumOD_{600}	Rate of nitrate removal (%)	Maximum NO-N accumulation (μg L^{-1})	Maximum N_2O-N accumulation (mg L^{-1})
Carbon source	Sodium succinate	0.82 (0.01) a	100.00 (0.00) a	3.20 (0.28) a	0.29 (0.09) a
	Sodium citrate	1.01 (0.01) b	100.00 (0.00) a	3.07 (0.28) a	2.63 (0.15) b
	Sodium acetate	0.77 (0.01) c	100.00 (0.00) a	2.61 (0.37) a	0.22 (0.11) a
	Glucose	0.47 (0.02) d	76.78 (2.93) b	85.46 (0.18) b	3.35 (0.18) c
	Sucrose	0.24 (0.00) e	10.94 (0.55) c	0.00 (0.00) c	0.04 (0.01) a
pH	6	0.23 (0.00) a	3.97 (0.95) a	0.59 (0.28) a	0.01 (0.00) a
	7	0.85 (0.02) b	99.20 (1.13) b	8.17 (0.65) b	0.01 (0.00) a
	8	0.88 (0.01) b	100.00 (0.00) b	6.01 (1.29) bc	0.04 (0.01) a
	9	0.87 (0.00) b	100.00 (0.00) b	4.51 (0.28) c	0.08 (0.01) a
	10	0.86 (0.00) b	100.00 (0.00) b	4.44 (0.18) c	0.18 (0.05) b
Rate of C/N	2	0.77 (0.01) a	88.67 (0.22) a	4.90 (0.09) a	0.58 (0.04) a
	4	0.93 (0.02) b	100.00 (0.00) b	4.64 (1.57) a	0.06 (0.00) b
	6	0.91 (0.00) bc	100.00 (0.00) b	4.25 (0.09) a	0.05 (0.00) b
	10	0.88 (0.01) cd	100.00 (0.00) b	6.01 (1.29) a	0.04 (0.01) b
	15	0.85 (0.01) d	100.00 (0.00) b	4.77 (1.20) a	0.04 (0.00) b

*The values in parentheses represent the standard error, and the same letters in the same column do not represent significant differences ($P < 0.05$)

disproportionation of HNO_2 (Zweier et al. 1995). The strain PCN-1 could not be used as a carbon source to grow or denitrify. Thus, sodium succinate and sodium acetate were suitable carbon sources for strain PCN-1, and sodium acetate was used as a carbon source in subsequent experiments in view of practical application.

It can be seen from Fig. 3.6 and Table 3.2 that strain PCN-1 could not grow or denitrify under conditions of acid. No significant difference was found in the growth and nitrification of PCN-1 at the initial pH 7–10 ($P > 0.05$). However, the production of N_2O at pH 10 was significantly higher than that at pH 7–9 ($P < 0.05$), but there

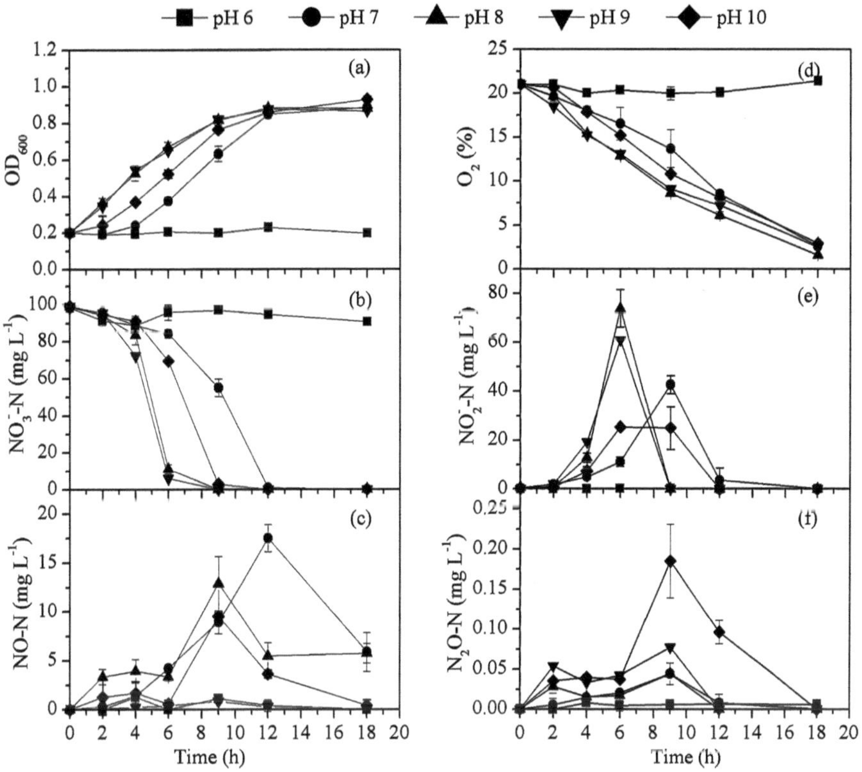

Fig. 3.6 Aerobic denitrification characteristics of strain PCN-1 under different pH values. (Reprinted from Zheng et al. (2014), with permission from Elsevier)

was no significant difference in the amount of NO produced at pH 7–9 ($P > 0.05$). Thus, neutral or weakly alkaline was a suitable condition for the growth and denitrification of the strain PCN-1.

It can be seen from Fig. 3.7 and Table 3.2 that under the condition of C/N = 2, the carbon source becomes the limiting factor of the growth and denitrification process of the strain PCN-1, so the denitrification process has accumulated a large amount of nitrite nitrogen (79.91 mg L^{-1}) and more N_2O-N (0.58 mg L^{-1}). When C/N rose to 4, nitrate and nitrite were all consumed, and only a little amount of NO and N_2O were accumulated. No significant difference in denitrification and gaseous production accumulation was observed when the C/N ratio was further increased to 6 or 15. Although it was considered that the C/N ratio of 8 or 20 was the suitable C/N value for aerobic denitrifying bacteria, strain PCN-1 could meet the requirement of denitrification for carbon source at lower C/N ratio, which greatly improved the potential of its application in the actual wastewater treatment.

In conclusion, carbon source species, value of pH, and ratio of C/N had significant effects on the growth, aerobic denitrification, and NO and N_2O production of

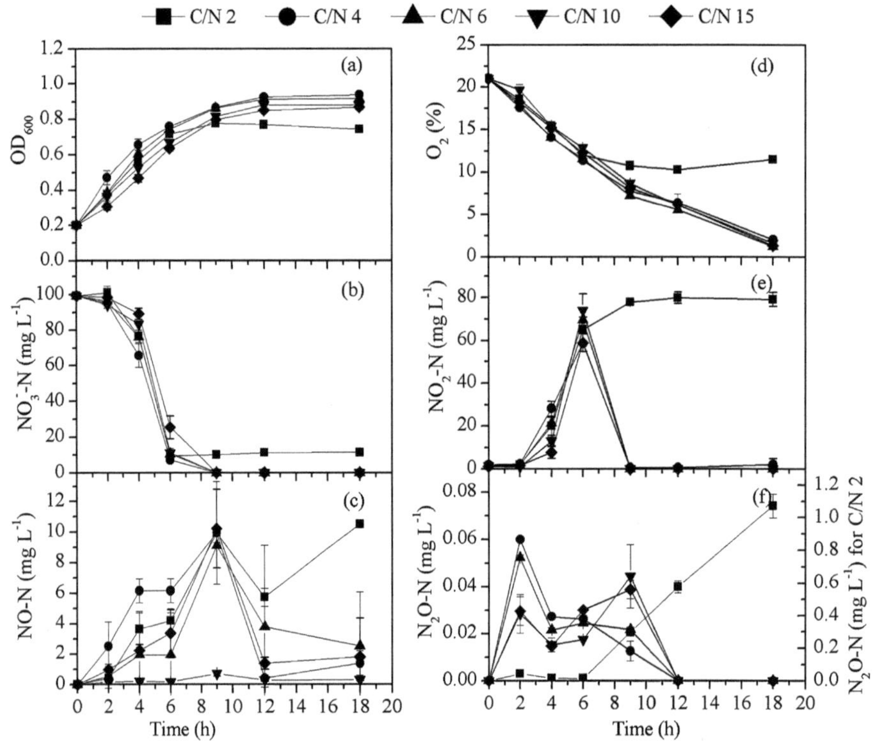

Fig. 3.7 Aerobic denitrification characteristics of strain PCN-1 under different C/N ratios

strain PCN-1. The strain PCN-1 grew well under the conditions of sodium succinate or sodium acetate as carbon source, the value of pH is 7 to 9, and the value of C/N was 4 to 6.

3.3.6 Strain PCN-1 on the Activated Sludge Enhanced Shake Flask Experiment

In view of its excellent denitrification performance and extremely low accumulation of NO and N_2O, the enhancement of the strain PCN-1 in the activated sludge system was further investigated. When there was only one nitrogen source of ammonia nitrogen in the simulated wastewater, the total nitrogen removal efficiency of the activated sludge system which inoculated with the strain PCN-1 was significantly increased, and no significant accumulation of nitrite was observed during the process. The maximum accumulation of nitrite was 2.06 mg L^{-1} in the contrast group (Fig. 3.8a), since nitrite accumulation was one of the main factors which lead to emission of NO and N_2O during biological denitrification (Alinsafi et al. 2008;

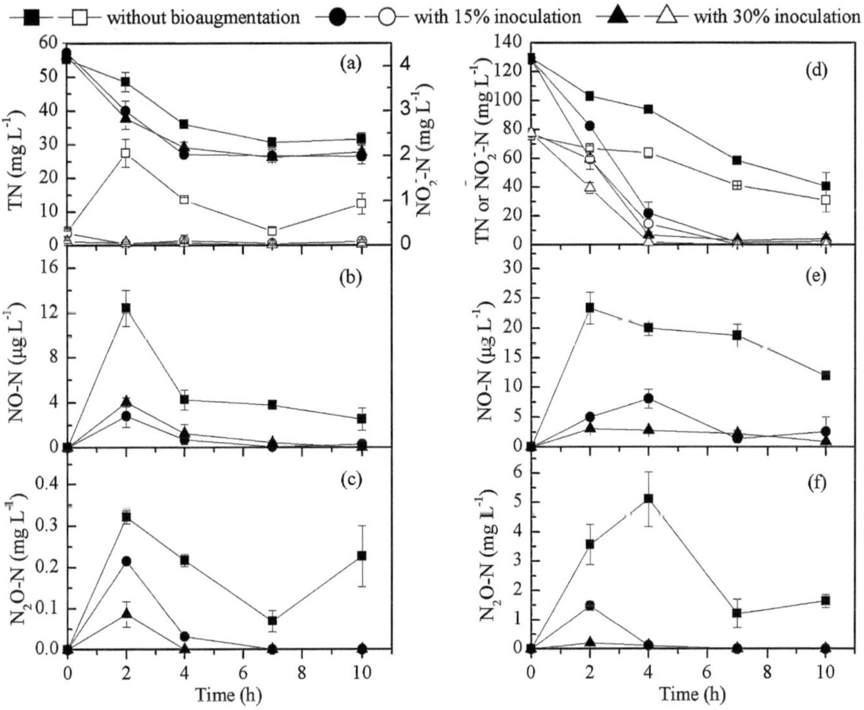

Fig. 3.8 Bioaugmentation of strain PCN-1 when nitrite was the main nitrogen pollutant. (Reprinted from Zheng et al. (2014), with permission from Elsevier)

Wunderlin et al. 2012). Therefore, rapid utilization of nitrite by strain PCN-1 was one of the important reasons for reduction of NO and N_2O emission (Fig. 3.8b and c). The amount of inoculation was an important factor that affected the enhancement effect of sludge. When the amount of inoculum increased from 15% to 30%, the emission of N_2O was reduced ($P < 0.05$) even though there was no significant change in the aspect of total nitrogen and NO ($P > 0.05$). Because the strain PCN-1 can be denitrified with exogenous N_2O as substrate, N_2O emitted by other microorganisms in the sludge could be directly reduced to N_2 by the strain PCN-1, which is another important reason for N_2O emission reduction by bioaugmentation.

When a large amount of nitrite was present in the simulated wastewater, strain PCN-1 still had a significant enhancement effect on the activated sludge system (Fig. 3.8d). As the inhibition of oxygen, nitrite degradation in the control group showed a significant lag phase, while nitrite nitrogen was quickly utilized in the experimental group. The final total nitrogen rate of removal reached 97.65% and 96.65% under the condition of 15% and 30% inoculation, respectively, which was much higher than 68.60% in the control group. At the same time, the emissions of NO and N_2O were also significantly reduced after inoculation with strain PCN-1 ($P < 0.05$, Fig. 3.8e and f).

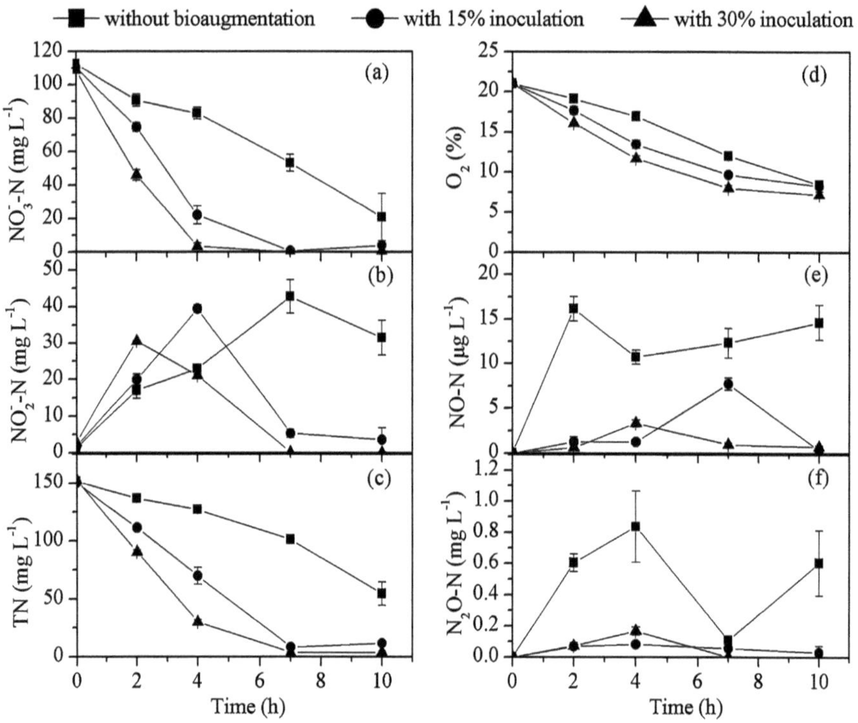

Fig. 3.9 Bioaugmentation of strain PCN-1 when nitrate was the main nitrogen pollutant. (Reprinted from Zheng et al. (2014), with permission from Elsevier)

When a large amount of nitrite nitrogen was present in the simulated wastewater, the denitrification rate in the activated sludge system inoculated with strain PCN-1 was significantly increased (Fig. 3.9). When the concentration of inoculum was 15% and 30%, the removal rate of nitric nitrogen was 79.96% and 96.82% at the initial concentration of 120 mg L^{-1} compared with 26.19% in the control group without PCN-1. In addition, the concentration of nitrite in the experimental group decreased rapidly after reaching the maximum value. The TN removal rate of the two experimental groups reached 94.47% and 97.70% in 7 h, respectively, compared with 33.18% in the control group, which showed a good denitrification effect of enhancement. It was similar to the results of the first two groups; the rapid application of nitrite and nitrogen oxides also resulted in the accumulation of NO and N_2O in the experimental group, which was significantly lower than that in the control group, and it showed a good emission reduction effect of nitrogen oxides.

It could be seen that the activated sludge system enhanced by PCN-1 not only had the efficiency and significant improvement on denitrification but also decreased the emission of NO and N_2O from the process of denitrification. Therefore, the strain PCN-1 had a great potential value of application in the activated sludge system of the removal of nitrogen by biological enhancement and the reduction of nitrogen oxide emission.

3.4 Summary

In this chapter, the aerobic denitrification performance of strain PCN-1 was systematically investigated, and the conclusions were drawn as follows:

1. The denitrification rate of PCN-1 was 11.66 mg L^{-1} h^{-1}, 12.80 mg L^{-1} h^{-1}, and 9.67 mg L^{-1} h^{-1} when the nitrification, nitrite, and N_2O as different nitrogen sources under the condition of aerobic, respectively. The maximum accumulation of NO and N_2O was only 0.003% and 0.33% of the total removal of nitrogen and nitrate and 0.006% and 0.29% of the total removal of nitrite. The co-expression of denitrifying genes nirS, cnorB, and nosZ under aerobic conditions was the main reason for the unique aerobic denitrification function of strain PCN-1.
2. The strain PCN-1 could degrade nitrate under the condition of O_2 concentration of 0–100% and show strong tolerance to high concentration of O_2. The accumulation of NO was significantly decreased with the increase of O_2 concentration, while the accumulation of N_2O increased significantly with the increase of O_2 concentration. The strain PCN-1 could degrade nitro-nitrogen under the condition of O_2 concentration 0–100% and showed strong tolerance to high concentration of O_2. The accumulation of NO was significantly decreased with the increase of O_2 concentration, while the accumulation of N_2O increased significantly with the increase of O_2 concentration. However, under the condition of 50% O_2 concentration, the accumulation of N_2O only accounted for 1.12% of the removal of nitrate.
3. The strain PCN-1 grew well, and the denitrification capacity was high, and the accumulation of NO and N_2O was low under the conditions of sodium succinate or sodium acetate as the carbon source, pH 7–9, and C/N 4–6, indicating that the strain PCN-1 was suitable for sewage treatment of low C/N ratio.
4. The bioremediation of strain PCN-1 enhanced the denitrification capacity of activated sludge system and significantly reduced the NO and N_2O emissions when the simulated wastewater was treated as the main pollutants with ammonia nitrogen, nitrite nitrogen, or nitric nitrogen. Thus it had a great potential application value for the reduction of nitrogen oxide emission in sewage treatment plants.

References

Alinsafi, A., Adouani, N., Beline, F., Lendormi, T., Limousy, L., & Sire, O. (2008). Nitrite effect on nitrous oxide emission from denitrifying activated sludge. *Process Biochemistry, 43*, 683–689.

Arts, P. A. M., Robertson, L. A., & Kuenen, J. G. (1995). Nitrification and denitrification by Thisphaera pantotropha in aerobic chemostat cultures. *FEMS Microbiology Ecology, 18*, 305–316.

Bergaust, L., Shapleigh, J., Frostegard, A., & Bakken, L. (2008). Transcription and activities of NO_x reductases in Agrobacterium tumefaciens: The influence of nitrate, nitrite and oxygen availability. *Environmental Microbiology, 10*, 3070–3081.

Cabiscol, E., Tamarit, J., & Ros, J. (2000). Oxidative stress in bacteria and protein damage by reactive oxygen species. *International Microbiology, 3*, 3–8.

Chen, J. W., & Strous, M. (2013). Denitrification and aerobic respiration, hybrid electron transport chains and co-evolution. *Biochimica et Biophysica Acta, 1827*, 136–144.

Chen, P. Z., Li, J., Li, Q. X., Wang, Y. C., Li, S. P., Ren, T. Z., & Wang, L. G. (2012). Simultaneous heterotrophic nitrification and aerobic denitrification by bacterium *Rhodococcus sp.* CPZ24. *Bioresource Technology, 116*, 266–270.

Guo, L. Y., Chen, Q. K., Fang, F., Hu, Z. X., Wu, J., Miao, A. J., Xiao, L., Chen, X. F., & Yang, L. Y. (2013). Application potential of a newly isolated indigenous aerobic denitrifier for nitrate and ammonium removal of eutrophic lake water. *Bioresource Technology, 142*, 45–51.

Hartig, E., & Zumft, W. G. (1999). Kinetics of *nirS* expression (cytochrome *cd1* nitrite reductase) in Pseudomonas stutzeri during the transition from aerobic respiration to denitrification: Evidence for a denitrification-specific nitrate- and nitrite-responsive regulatory system. *Journal of Bacteriology, 181*, 161–166.

Kester, R. A., De Boer, W., & Laanbroek, H. J. (1997). Production of NO and N_2O by pure cultures of nitrifying and denitrifying bacteria during changes in aeration. *Applied and Environmental Microbiology, 63*, 3872–3877.

Miyahara, M., Kim, S. W., Fushinobu, S., Takaki, K., Yamada, T., Watanabe, A., Miyauchi, K., & Endo, G. (2010). Potential of aerobic denitrification by Pseudomonas stutzeri TR2 to reduce nitrous oxide emissions from wastewater treatment plants. *Applied and Environmental Microbiology, 76*, 4619–4625.

Muyzer, G., De Waal, E. C., & Uitterlinden, A. G. (1993). Profiling of complex microbial populations by denaturing gradient gel electrophoresis analysis of polymerase chain reaction-amplified genes coding for 16S rRNA. *Applied and Environmental Microbiology, 59*, 695–700.

Otte, S., Grobben, N. G., Robertson, L. A., Jetten, M. S., & Kuenen, J. G. (1996). Nitrous oxide production by *Alcaligenes faecalis* under transient and dynamic aerobic and anaerobic conditions. *Applied and Environmental Microbiology, 62*, 2421–2426.

Philippot, L. (2002). Denitrifying genes in bacterial and archaeal genomes. *Biochimica et Biophysica Acta, 1577*, 355–376.

Pomowski, A., Zumft, W. G., Kroneck, P. M. H., & Einsle, O. (2011). N_2O binding at a [4Cu:2S] copper-sulphur cluster in nitrous oxide reductase. *Nature, 477*, 234–237.

Poole, R. K., & Hughes, M. N. (2000). New functions for the ancient globin family: Bacterial responses to nitric oxide and nitrosative stress. *Molecular Microbiology, 36*, 775–783.

Throback, I. N., Enwall, K., Jarvis, A., & Hallin, S. (2004). Reassessing PCR primers targeting nirS, nirK and nosZ genes for community surveys of denitrifying bacteria with DGGE. *FEMS Microbiology Ecology, 49*, 401–417.

Saleh-Lakha, S., Shannon, K. E., Goyer, C., Trevors, J. T., Zebarth, B. J., & Burton, D. L. (2008). Nitric oxide reductase gene expression and nitrous oxide production in nitrate-grown *Pseudomonas mandelii. Applied and Environmental Microbiology, 74*, 6876–6879.

Spiro, S. (2012). Nitrous oxide production and consumption: Regulation of gene expression by gas-sensitive transcription factors. *Philosophical Transactions of the Royal Society B, 367*, 1213–1225.

Wan, C. L., Yang, X., Lee, D. J., Du, M. A., Wan, F., & Chen, C. (2011). Aerobic denitrification by novel isolated strain using NO_2^--N as nitrogen source. *Bioresource Technology, 102*, 7244–7248.

Wunderlin, P., Mohn, J., Joss, A., Emmenegger, L., & Siegrist, H. (2012). Mechanisms of N_2O production in biological wastewater treatment under nitrifying and denitrifying conditions. *Water Research, 46*, 1027–1037.

Yang, X. P., Wang, S. M., Zhang, D. W., & Zhou, L. X. (2011). Isolation and nitrogen removal characteristics of an aerobic heterotrophic nitrifying-denitrifying bacterium, *Bacillus subtilis* A1. *Bioresource Technology, 102*, 854–862.

Yao, S., Ni, J. R., Ma, T., & Li, C. (2013). Heterotrophic nitrification and aerobic denitrification at low temperature by a newly isolated bacterium, *Acinetobacter sp.* HA2. *Bioresource Technology, 139*, 80–86.

Zhang, Q. L., Liu, Y., Ai, G. M., Miao, L. L., Zheng, H. Y., & Liu, Z. P. (2012a). The characteristics of a novel heterotrophic nitrification-aerobic denitrification bacterium, *Bacillus methylotrophicus* strain L7. *Bioresource Technology, 108*, 35–44.

Zhang, T., Shao, M. F., & Ye, L. (2012b). 454 pyrosequencing reveals bacterial diversity of activated sludge from 14 sewage treatment plants. *ISME Journal, 6*, 1137–1147.

Zheng, M. S., He, D., Ma, T., Chen, Q., Liu, S. T., Ahmad, M., Gui, M. Y., & Ni, J. R. (2014). Reducing NO and N_2O emission during aerobic denitrification by newly isolated Pseudomonas stutzeri PCN-1. *Bioresource Technology, 162*, 80–88.

Zumft, W. G. (1997). Cell biology and molecular basis of denitrification. *Microbiology and Molecular Biology Reviews, 61*, 533–616.

Zweier, J. L., Wang, P., Samouilov, A., & Kuppusamy, P. (1995). Enzyme-independent formation of nitric oxide in biological tissues. *Nature Medicine, 1*, 804–809.

Chapter 4
Application of *Pseudomonas stutzeri* PCN-1 in N_2O Emission Reduction

4.1 Introduction

Since the 1990s, aerobic denitrifying bacteria have been used in the enhanced biological denitrification process of activated sludge systems. It was found that the supplementation of aerobic denitrifying bacteria could simultaneously make biological denitrification system finish nitrification and denitrification with the advantages of high denitrification efficiency, no alkali neutralization, and strong anti-shock loading ability, manifesting the superiority over traditional method of wastewater treatment (Bouchez et al. 2009; Yang et al. 2011). With the popularization and enlargement of sewage biological treatment process, the contribution of N_2O emission to global greenhouse effect is increasing day by day. How to reduce the NO and N_2O emission from sewage denitrification process has become a research hotspot.

Oxidation ditch is widely used in sewage treatment plants because of its high denitrification efficiency and low running cost. This process combines the characteristics of a full-flow reactor and a plug flow reactor, which has a significant effect on N_2O emission characteristics. On one hand, the large internal reflux ratio (greater than 50 times) of the oxidation ditch system possess the characteristic of continuous stirred tank reactor, which maintained. The nitrogen concentration at a low level and reduced the driving force of N_2O generation (Foley et al. 2010). On the other hand, oxidation ditch has the characteristics of plug flow reactor, where activated sludge alternatively flow through the aerobic and anoxic area to complete the nitrification and denitrification, which in turn promotes the N_2O production.

In this chapter, the application potential of N_2O in denitrification process was explored on the basis of denitrifying SBR basing on the study of N_2O emission characteristics and mechanism of strain PCN-1, and then the process of N_2O was studied in the pilot-scale Carrousel oxidation ditch. The effect of PCN-1

M. Zheng, *Nitrogen Removal Characteristics of Aerobic Denitrifying Bacteria and Their Applications in Nitrogen Oxides Emission Mitigation*, Springer Theses,
https://doi.org/10.1007/978-981-13-2432-1_4

bio-fortification on the stable operation of the system and the emission of nitrogen oxides was explored after the emission characteristics and emission mechanism of N_2O were investigated.

4.2 Materials and Methods

4.2.1 Start-Up and Operation of Denitrifying SBR Reactors

Two parallel SBR reactors (R1 and R2) were set up to verify the N_2O emission reduction performance of the PCN-1 during the biological denitrification process. The reactor was made of plexiglass with an inner diameter of 15 cm, a height of 50 cm, and a volume of 5 L. The top of the reactor was connected to a plexiglass lid by a flange, and the reactor was sealed with a gasket. The reactor lid was provided with specific openings for the DO probe, pH probe, and gas and water sampling ports for monitoring and sampling analysis during biochemical reactions. During the reaction stage, the activated sludge was mixed with the simulated wastewater through an electric mixer and uniformly aerated by sand core gas stone placed flat at the bottom of the reactor.

The inoculation sludge was taken from the sewage treatment plant of Langfang Development Zone. The operation cycle was performed six times daily including 10 min water intake, 3 h reaction, 10 min aeration, 30 min standing, 5 min drainage, and 5 min idle. Per cycle started by a 2.5 L water injection through the peristaltic pump and stopped by a 2.5 L water discharge through the solenoid valve taking a hydraulic retention time of 8 h. After the end of the reaction, 10 min aeration was set to blow out the bubbles in the sludge flocs to assist the sludge settling. At the end of the reaction period, 500 mL of sludge was discharged every day, and the residence time of sludge was maintained for 10 days. The reactor was maintained at a constant temperature of 30 °C through a water bath and a water bath sandwich.

In this study, N_2O emission characteristics during denitrification were investigated by simulating wastewater treatment. The formula was that per liter, water contained 3.40 g $CH_3COONa \cdot 3H_2O$, 2.43 g $NaNO_3$, 0.35 g KH_2PO_4, 0.10 g $MgSO_4 \cdot 7H_2O$, 0.05 g $CaCl_2$, 0.003 g $FeSO_4 \cdot 7H_2O$, and 0.1 mL of trace element stock solution. The simulated wastewater COD was 1600 mg L^{-1}, the nitrate concentration was 400 mg L^{-1}, and the COD/N was 4:1. During the start-up phase, each cycle entered the hypoxia agitation reaction stage (without aeration) after water flowing in. By routine monitoring of the effluent water quality of the reactor, it was found that almost no nitrite and nitrate could be detected in the effluent. After the two sludge-age stable operations, the bioaugmentation experiment was started. The activated sludge concentration (MLSS) in the reactor was about 2600 mg L^{-1}.

4.2.2 Bioaugmentation of Denitrifying SBR

In order to study the bioaugmentation and N_2O emission reduction performance of PCN-1 under different dissolved oxygen conditions, the SBR system was operated under three different oxygen concentration conditions by controlling different nitrogen and air flow rates. In the first stage, nitrogen of high purity was introduced into the reactor, and the aeration rate was 100 mL min^{-1}. In the second stage, the mixed gas was injected into the reactor through a three-way valve, the flow rate of nitrogen was 75 mL min^{-1}, and the flow rate of air was 25 mL min^{-1} (O_2 content was about 5%). In the third stage, the mixed gas was exposed to the SBR through the three-way valve, in which the nitrogen flow rate was 50 mL min^{-1} and air flow was 50 mL min $^{-1}$ (O_2 content was about 10%).

The strain PCN-1 was pre-cultured to late logarithmic growth stage in LB medium. Then the cells were collected by centrifugation and inoculated into the reactor R2 at an inoculation amount of 15% (v/v), and an equal amount of the ordinary sludge was put into the reactor R1 as a blank control. The two reactors were exposed for 2 days without water in or out. And only a small amount of concentrated medium was added so that the bacteria could be effectively combined with the sludge in the reactor. After that, we returned to normal of the SBR cycle and monitored the two SBR out of the water quality. When the condition changed, the reactor must run at least one sludge age and do an analysis of gas and water samples to ensure the stability of the system.

In order to investigate the effect of PCN-1 on the bioremediation of denitrifying SBR, the water quality and gas samples of the typical cycle were sampled and analyzed periodically. Specifically, 5 mL of the sludge mixture was withdrawn from the sampling port every 30 min. After centrifugation, the supernatant was stored for determination of the nitrogen concentration. The exhaust gas was directly introduced into the gas chromatograph (GC) and NO through the exhaust port at the top of the reactor. The N_2O and NO concentrations in the exhaust gas were collected and analyzed online every 5 min. The dissolved oxygen and pH probes were used to monitor the DO and pH changes of the system in real time.

4.2.3 Start-Up and Operation of Oxidation Ditch

The experimental device was a pilot-scale Carrousel oxidation ditch in this study. The reactor was made of a plexiglass material consisting of a 50 L bio-selecting tank and four channels of 1.15 m long, 0.35 m wide, and 0.5 m deep. Its working volume was 1.4 m^3 (Fig. 4.1). First the inlet and the return sludge entered the selection tank and then entered the oxidation ditch from the first ditch. The sludge flowed in the mixed liquid ditch, and the outlet was set at the end of the fourth ditch. Two surface turbines and four submerged agitators were installed along the ditch to facilitate the flow of sludge mixture at a rate of 0.01 m s^{-1}. The effluent entered a sedimentation

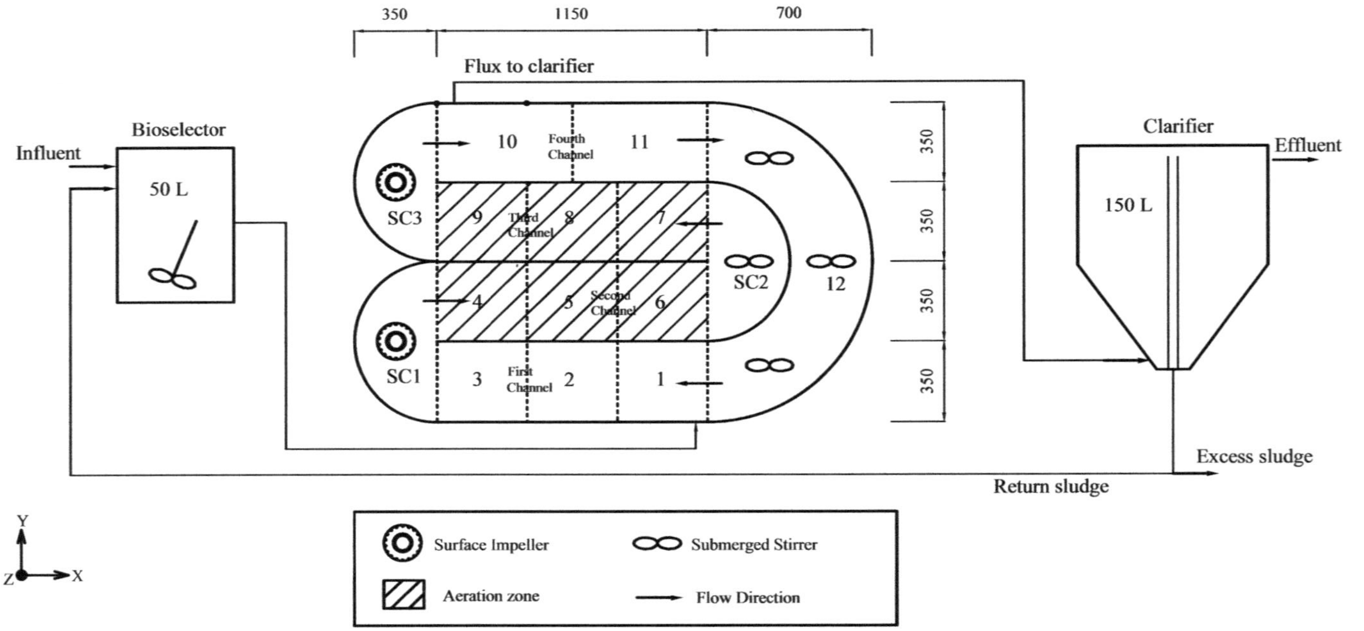

Fig. 4.1 The sketch of the pilot-scale Carrousel oxidation ditch and sampling zones. (Reprinted from Zheng et al. (2015), with permission from Elsevier)

tank with a volume of 0.15 m^3 for sludge separation and returned the sludge to the selection tank with a sludge reflux ratio of 100%. In order to study the spatial distribution of N_2O emission in the oxidation ditch systematically, the channel was divided into 12 independent zones to facilitate the collection of gas and water samples. In total, 14 sampling points were set in the biological selection tank and sedimentation tank.

Sludge in oxidation ditch was taken from the sewage treatment plant in Langfang Development Zone. The simulated water distribution was stored in a 1.8 m^3 water storage tank. The water was pumped into the selection tank at a flow rate of 0.1 m^3 h^{-1} through a metering pump to make the hydraulic retention time (HRT) maintained at 14 h. Each simulated water solution contained 531.9 g $CH_3COONa{\bullet}3H_2O$, 191.1 g NH_4Cl, 36.8 g $K_2HPO_4{\bullet}3H_2O$, 12.0 g $MgSO_4{\bullet}7H_2O$, 10.0 g $CaCl_2$, 3.0 g $FeSO_4{\bullet}7H_2O$, and 50 ml of trace element mother liquor. COD was 250 mg L^{-1}, and the ammonia nitrogen concentration was 50 mg L^{-1}.

According to the experience of the previous reactor operation, the aeration flow was controlled at 1.8 m^3 h^{-1} through the flow meter. The aeration was carried out by four 1-m-long microporous aeration pipes provided at the bottom of the second and third trenches. 60 L sludge mixture was discharged from the fourth channel every day, and the sludge retention time (SRT) of the oxidation ditch was maintained at about 25 days. The temperature in the oxidation ditch was controlled at 22 $\pm$ 2 °C by means of a temperature-controlled heating rod, and the pH was not adjusted (about 7.8). COD, NH_3-N, NO_2-N, and NO_3-N concentrations were detected daily for the influent and effluent of oxidation ditch. Two months after the stable operation of the oxidation ditch system, 14 samples in the oxidation ditch were used to collect the tail gas from the tank, and the mud and water mixture in the corresponding area was collected. The N_2O emission characteristics of the oxidation ditch system were investigated weekly in detail for 5 weeks.

4.2.4 Biological Enhancement of the Oxidation Ditch System

It was found that N_2O emission factor can be maintained at a low level when the system was operated at a relatively high sludge age through the analysis of the long-term operation of the oxidation ditch and its N_2O emission characteristics. When the system was operated at a low sludge age or subjected to a load shock, the N_2O emission increased significantly, which was mainly due to the reduction of microbial abundance and the enhancement of autotrophic denitrification process due to the accumulation of nitrite (Lotito et al. 2012). In order to investigate the effect of strain PCN-1 on enhancing the stable operation capability of the activated sludge system and the N_2O emission reduction, the exhaust gas from the six aeration zones was collected and analyzed after completion of the N_2O emission characteristics of the system, and the sludge age of the oxidation ditch system was controlled at 17 days (two sludge ages) to reach steady-state again. The same collection was carried out once a week for 5 weeks.

In order to study the change of N_2O emission in the oxidation ditch system under abnormal interference, two different impact tests were carried out on the oxidation ditch by manual intervention after the completion of the sampling and analysis of the five aeration zones, namely, high ammonia load shock and aeration abnormal shock. In the high ammonia load impact experiment, the influent NH_3-N was increased from 50 mg L^{-1} to 250 mg L^{-1} instantaneously. The remaining operating conditions were kept constant and then returned to normal water supply after 1 h of continuous high NH_4^+ influent. During the aeration experiment, the aeration flow rate was reduced from 1.8 m^3 h^{-1} to 0.9 m^3 h^{-1} for 2 h and then returned to normal aeration. In the impact test, the exhaust gas of the zone 5 and the zone 8 was collected at intervals of 10–30 min to represent the exhaust emissions of the two aeration zones, and the DO dynamic change was measured in the zone 4 and the zone 9. At the same time, 10 mL of mud and water mixture was taken after centrifugation analysis of nitrogen concentration in water changes.

In order to make the microbe work better in the system, the strain PCN-1 was immobilized and then added to the oxidation ditch. The strain PCN-1 was grown in a shake flask to a logarithmic growth phase and then inoculated in a 200 L culture tank filled with a polyurethane foam carrier having a volume of about 120 L. The temperature was maintained at 30 °C by a heating rod, and the medium was changed every 2 days to make the bacteria grow adhering to the carrier. After a period of 10 days, the carrier was added into the oxidation ditch, about 30 L of which was added to the selection tank, and the remaining 90 L of the culture medium was added to 4 channels under the flow of water to promote circulation. The effects of N_2O emission in the aeration zone on the N_2O load and the impact of aeration on the N_2O emission in the aeration zone were investigated and compared with the results before the bioremediation.

4.2.5 Gas Collection and N_2O Emission Rate Calculation

Sampling methods and N_2O emission rates for different zones of the oxidation ditch are described below:

1. Aeration zone

The tail gas in the aeration zone was collected by floating tank method. The float chamber size was 23.5 cm × 18.5 cm × 10 cm. After the float chamber was fixed to the sampling area, 6.6 cm was submerged under the liquid surface. A 2 L air pocket was connected to the pontoon outlet to collect the gas directly to make surrounding surface area at 434.75 cm^2; the space of pontoon above water was 1.48 L. The N_2O emission rate was calculated as shown in Eq. (4.1):

$$F_{i-aero} = C_{i-aero} \times Q_T / A_T \tag{4.1}$$

Among which, $F_{i\text{-aero}}$ represented the emission rate of region i (μg m^{-2} min^{-1}); Ci-aero represented the N_2O-N mass concentration in the off-gas of region i (μg L^{-1}); the conversion relationship between N_2O volume concentration and mass concentration is that 1 ppm N_2O = 1.13 μg L^{-1} N_2O-N at 20 °C; QT represented the total aeration rate (L min^{-1}); and AT represented the total area of the aeration area (m^2).

2. Non-aeration zone

Because of the absence of obvious gas discharge in the aeration zone, the floating tank was suspended in the liquid surface for 20 min, and 20 mL of the air was taken from the bioreactor and the sedimentation tank every 5 min. The N_2O emission rate in the non-aeration zone was calculated based on the linear variation of the N_2O concentration in the pontoon, the surface area enclosed by the pontoon, and the space volume on the surface (Beaulieu et al. 2010). The calculation formula is shown in (4.2).

$$F_{i-nonaero} = (dC_{i-nonaero}/dt) \times V/A \tag{4.2}$$

Among which, Fi-nonaero (μg m^{-2} min^{-1}) is the rate of N_2O emission in the non-aeration zone; i is 1–3 and 10–12, biological selection tank and sedimentation tank; $dC_{i\text{-nonaero}}/dt$ is the slope of the N_2O concentration linear; V is the volume above the float tank surface (L); and A was the liquid surface area surrounded by float box (m^2).

3. The dissolved state

Dissolved N_2O in the mixed solution of each zone was measured by the upper-space method (Yang et al. 2009): 25 mL of the mixture was removed from 10 cm below the surface of the sampling area with a 50 mL gas-tight syringe and immediately injected with 1 mL of 3 g L^{-1} $HgCl_2$ solution, and then 25 mL of high-purity nitrogen was extracted; the syringe was sealed and violently shocked for 2 min and then put aside for 1 h until the gas-liquid equilibrium in the syringe determines the gas phase N_2O concentration. As a result, dissolved N_2O concentrations were calculated using Henry's coefficient (Weiss and Price 1984) as shown in Eq. 4.3.

$$C_{i-diss} = P_i \times K_0 \tag{4.3}$$

Among which, $C_{i\text{-diss}}$ represented the concentration of dissolved N_2O (μg L^{-1}, i = 1–12), Pi represented the equilibrium partial pressure of N_2O in the upper space of the syringe (atm), and K_0 is the volume solubility coefficient of N_2O at 22 °C, $K_0 = 7.578 \times 10^5$ μg L^{-1} atm^{-1}.

4. Total gaseous emissions

The N_2O emission factor of each region was calculated from the emission rate, the area, the inflow rate, and the NH_3-N concentration. The total N_2O emission factor of the whole oxidation ditch system was the sum of the N_2O emission factors of each region, as shown in Eqs. 4.4 and 4.5.

$$EF_i = (F_i \times A_i)/(C_{NH4-N} \times Q_{inf} \times 1000) \times 100\% \quad (4.4)$$

$$EF_{gas} = \sum EF_i \quad (4.5)$$

where EFi is the emission factor for region I (%, i = 1–12), EF_{gas} is the total N_2O emission factor of oxidation ditch (%), Fi is N_2O emission rate in the area i (μg m^{-2} min^{-1}), Ai is the surface area of the area i (m^{-2}), $C_{NH4\text{-}N}$ is the influent ammonia concentration (mg L^{-1}), and Q_{inf} is the influent flow rate (L min^{-1}).

5. Total liquid emissions

Dissolved N_2O with drainage eventually entered into the environmental water, so the liquid emissions were also calculated in the total emission factor of the system, which was calculated as shown in Eq. 4.6.

$$EF_{liq} = C_{cla}/C_{NH4-N} \quad (4.6)$$

where EF_{liq} represented the emission factor of dissolved N_2O (%) and $C_{NH4\text{-}N}$ represented the influent ammonia concentration (mg L^{-1}).

6. Total emissions

The total N_2O emissions from the oxidation ditch system are the sum of the total gaseous and liquid emissions, as shown in Eq. 4.7.

$$EF = EF_{gas} + EF_{liq} \quad (4.7)$$

where EFliq is the total N_2O emission factor of the system (%).

4.2.6 N_2O Generation Mechanism in Oxidation Ditch

In order to investigate the contribution of heterotrophic denitrification (HD) and autotrophic denitrification (ND) to N_2O emission in oxidation ditch, batch experiments were carried out in a closed fermentation tank. The effective volume of the fermentation tank was 3.5 and 1.5 L of the upper space. The mixture was removed from the fourth furrow of the oxidation ditch, and aeration was carried out for 6 h to remove the remaining organic matter. The sludge was washed twice and then added to the reactor for 2 h (Wunderlin et al. 2012). The concentration of dissolved oxygen in the reactor was controlled at 1.0–2.0 mg L^{-1}, and the concentration of dissolved oxygen in the reactor was controlled at 1.0–2.0 mg L^{-1}, respectively. In the experiment, five DO gradients were set up, 1.0–2.0 mg L^{-1}, 0.5–1.0 mg L^{-1}, 0.2–0.5 mg L^{-1}, 0.05–0.2 mg L^{-1}, and 0–0.01 mg L^{-1} (strict hypoxia), to simulate dissolved oxygen conditions in different zones of the oxidation ditch. The aeration flow was maintained at 0.4 L min^{-1}. To ensure better control conditions, air was used in place of high-purity oxygen at O_2 concentrations of 0.05–0.2 mg L^{-1} and in severe hypoxic conditions. N_2O and tri-N concentrations were analyzed at regular intervals from the reactor using a mixture of gases and mixtures.

In order to study the denitrification rate and N_2O production rate of different biochemical processes under different DO conditions, four batches of experiments were carried out under each DO condition. The environment in the experimental reactor was as follows:

1. With HD and ND simultaneously: N_2O production rate was COD 125 mg L^{-1}, ammonia nitrogen 25 mg L^{-1}, and nitrite nitrogen 10 mg L^{-1}.
2. HD rate and N_2O emission generation rate: in addition to (1), 10 mg L^{-1} ATU and 1 g L^{-1} $NaClO_3$.
3. The rate of nitrite oxidation and its rate of N_2O production: nitrite nitrogen was 25 mg L^{-1}.
4. Nitrite nitrogen oxidation rate and its N_2O production rate.

The experiment was terminated when no significant N_2O production was observed during the course of the reaction or when the ammonia was completely converted to nitric nitrogen. The N_2O production rate of the autotrophic denitrification process was calculated by subtracting the N_2O production rate (2) from the total production rate (1) from the heterotrophic denitrification process (Tallec et al. 2006).

4.2.7 Analysis of Sludge Community Structure Based on High-Throughput Sequencing

Activated sludge samples were collected in a stable oxidation ditch, and total genomic DNA was extracted from the sludge using the Soil DNA Rapid Extraction Kit (MP Biomedical, Illkirch, France). The V4 region of 16S rRNA was amplified with primers 515F (GTGCCAGCMGCCGCGGTAA) and 806R (GGACTACHVGGGTWTCTAAT). The amplification conditions were as follows: 95 °C denaturation for 2 min, 95 °C denaturation for 30 s; 53 °C annealing for 20 s; 72 °C extension for 60 s, a total of 30 cycles; and 72 °C extension for 5 min. After confirmation by agarose gel electrophoresis, the PCR products were sequenced on the Illumina Miseq platform.

Sequencing results were performed using QIIME v.1.7.0 for quality control, noise reduction, and chimer detection. The remaining 28,579 sequences were then grouped into 1035 different OUTs at 97% similarity level using the UPARSE software, followed by aligning in the Greengenes "gg_13_8" database for species annotation using UCLUST v1.2.22q software (DeSantis et al. 2006). The sequencing results were submitted to the National Center for Biotechnology Information (NCBI), a small fragment sequence set (SRA) database under accession number SRP048693.

4.2.8 Analytical Methods

After centrifugation, the concentration of ammonia nitrogen, nitrite, and nitrate in the supernatant was analyzed by automatic chemical analyzer (Smartchem 200, AMS, Italy). COD was determined by titration, and TOC was measured using a TOC/TN analyzer (Multi N/C 3000, Analytik Jena, Germany). Both PHA and N_2O were detected by a gas chromatograph (GC) equipped with a flame particle detector (FID) and an electron capture detector (ECD). The statistical differences in nitrogen concentration and N_2O emission rates in different sampling areas were statistically different from ANOVA and nonparametric Friedman tests when $P < 0.05$.

4.3 Results and Discussion

4.3.1 Bioaugmentation Effect of Denitrifying SBR

Figures 4.2 and 4.3 showed the changes of nitrogen concentration, DO, and pH in two typical SBR cycles under hypoxic conditions. At the start of the cycle, the DO in

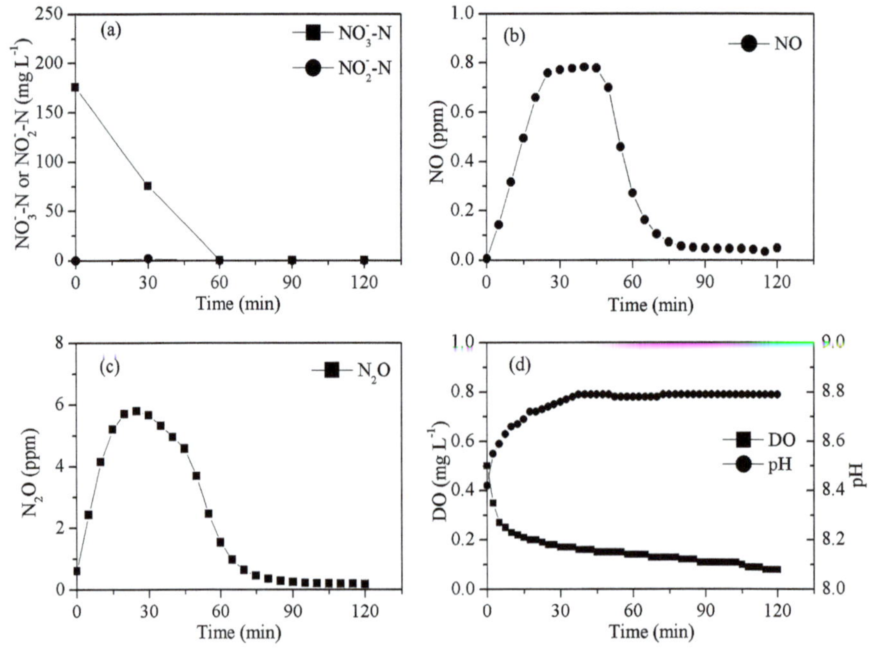

Fig. 4.2 Variations of -N and -N concentration (**a**), NO (**b**) and N_2O (**c**) emission rates, DO and pH (**d**) in R1 under anoxic condition

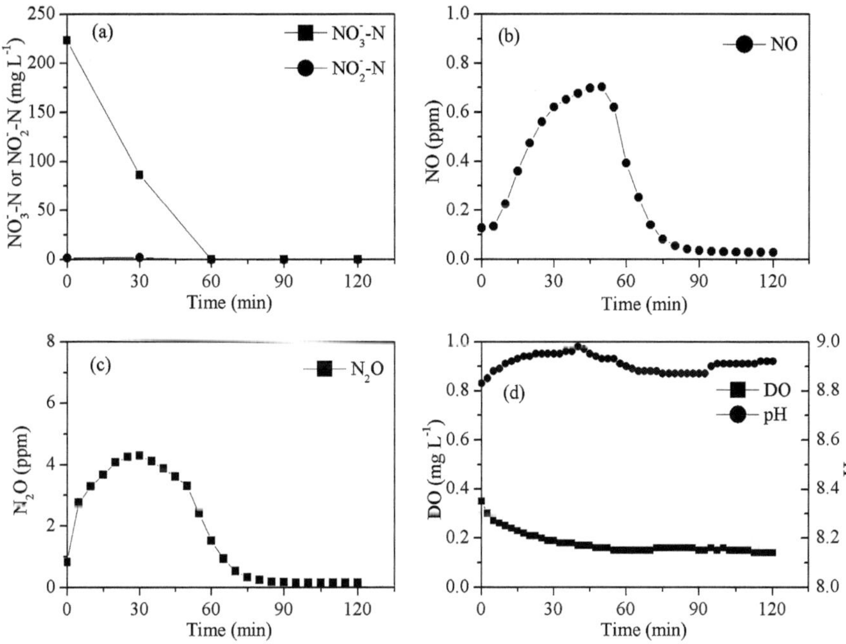

Fig. 4.3 Variations of -N and -N concentration (**a**), NO (**b**) and N_2O (**c**) emission rates, DO and pH (**d**) in R2 under anoxic condition

the SBR was about 0.4 mg L^{-1}, which is due to the DO brought in by the influent. With the absence of oxygen in the aeration, the DO is rapidly reduced to less than 0.1 mg L^{-1} and continued to decrease throughout the reaction period. In the absence of oxygen, the removal rate of nitrate nitrogen in the two SBRs was fast. The nitrate nitrogen of 200 mg L^{-1} was degraded within 60 min in the influent, and there was almost no accumulation of nitrite during the nitrification process. The pH value of the bioremediation reactor R2 was higher in the whole process, which may be related to the difference of metabolic function of different microbial communities because of the acidity in the process of denitrification.

NO and N_2O emissions occurred mainly in the process of removal of nitrification, and emission concentration showed a typical bell-shaped trend. By comparing the trend of N_2O and NO_3-N, it could be seen that the inflection point of N_2O emission rate appeared in the first 30 min, and higher concentration of nitric nitrogen was still present, probably due to the dissolved oxygen in the initial influent N_2O reductase inhibition. By comparing the N_2O emission concentrations of the two SBR reactors, it was found that the N_2O emission concentration of the bio-reactivated R2 reactor was significantly reduced and the total calculated emission fraction was reduced by 19.81%, indicating that the strain PCN-1 in the activated sludge system and the N_2O reduction effect was exerted well.

Figures 4.4 and 4.5 show the variation of SBR denitrification during the aeration of 5% oxygen concentration. The initial DO value was 0.6 mg L^{-1}, and the

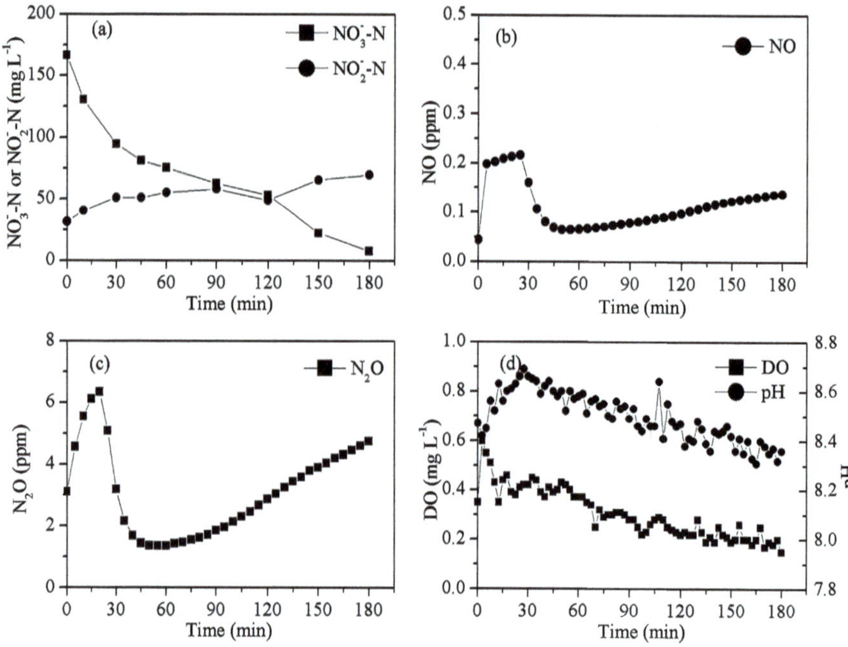

Fig. 4.4 Variations of -N and -N concentration (**a**), NO (**b**) and N_2O (**c**) emission rates, DO and pH (**d**) in R1 under aeration of 5% O_2

denitrification rate decreased obviously. Compared with the anaerobic-anaerobic condition, the nitrification was not required after 180 min, and the accumulation of nitrite was continued during denitrification. The final accumulation was as high as 70 mg L^{-1}. In the bioremediation of R2, although the denitrification rate decreased compared with the hypoxia conditions, nitride nitrogen would still be fully degraded at 90 min in the water. Despite the highest 28 mg L^{-1} of the subnitrate accumulation, almost all nitrite nitrogen degraded in 120 min. The results showed that the aerobic denitrification capacity of SBR was enhanced by PCN-1 bio-fortification, and the aerobic denitrification capacity was still strong.

From the analysis of gaseous products, both NO and N_2O emissions occurred in the first 45 min, and this stage was also the fastest stage of nitrification degradation. Similar to the hypoxic conditions, NO release from the control reactor SBR1 also experienced a plateau. After 45 min, the denitrification process in R1 was inhibited by oxygen, and the concentration of NO_2-N increased, and the production of NO and N_2O began to increase gradually. The presence of aerobic denitrifying bacteria PCN-1 in R2 could maintain the denitrification process even under the condition of DO of 0.5 mg L^{-1}. With the accumulation of nitrite nitrogen gradually degraded, NO and N_2O also showed a continuous downward trend. After calculation, the total N_2O emission in single cycle after bioremediation decreased by 46.12% compared with the control group.

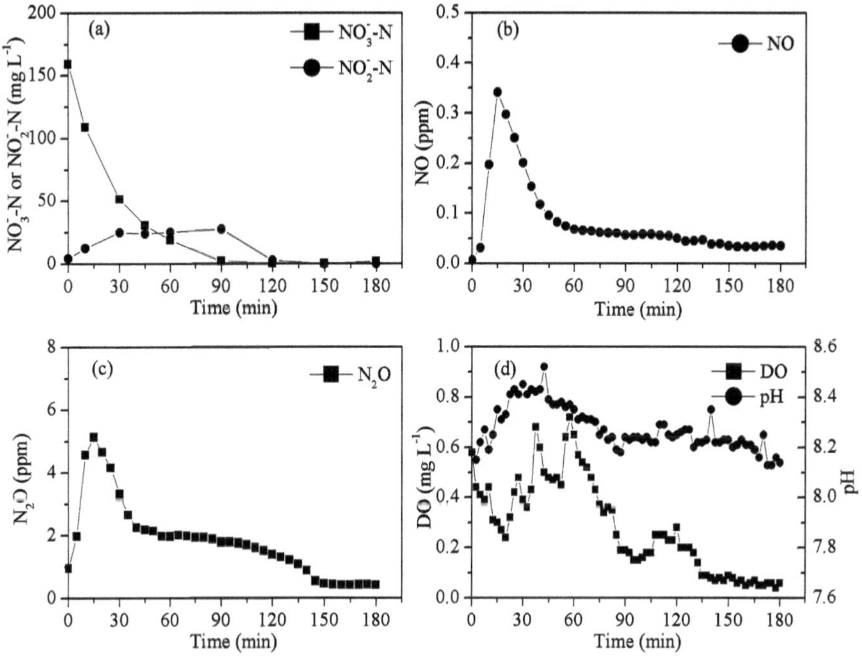

Fig. 4.5 Variations of -N and -N concentration (**a**), NO (**b**) and N_2O (**c**) emission rates, DO and pH (**d**) in R2 under aeration of 5% O_2

Nitrification removal at 10% oxygen concentration in aeration was similar to that at 5% (Figs. 4.6 and 4.7), and the proportion of organic matter assimilated was increased due to the presence of oxygen, leading to insufficient denitrification. However this condition improved significantly after bioremediation. At 120 min, nitrite and nitrosamines were completely denitrified, which was due to the fact that strain PCN-1 could still exert high denitrification under high dissolved oxygen condition. The NO emission in both SBRs was lower than that in the first two dissolved oxygen concentrations due to the enhanced oxidation. Unexpectedly, N_2O emissions were significantly lower in the first two dissolved oxygen conditions, even in the non-bio-fortified R1, probably due to the long-term acclimation of the sludge to higher dissolved oxygen concentrations, forming a more compact sludge floc, so that the N_2O reduction can be carried out smoothly. Even so, the bio-fortified SBR was still 45.13% less than the control.

Table 4.1 summarized the effects of SBR single cycle NO_x reduction before and after bioaugmentation. It can be seen that the bioremediation of strain PCN-1 could significantly reduce nitrogen oxides emission during SBR denitrification. Under the O_2 concentration of 0%, 5%, and 10%, the N_2O emission from PCN-1 after bio-fortification under aerobic conditions (19.81%, 46.12%, and 45.13%) was better than that under hypoxic conditions, which was consistent with the aerobic denitrification characteristics of strain PCN-1 under aerobic conditions.

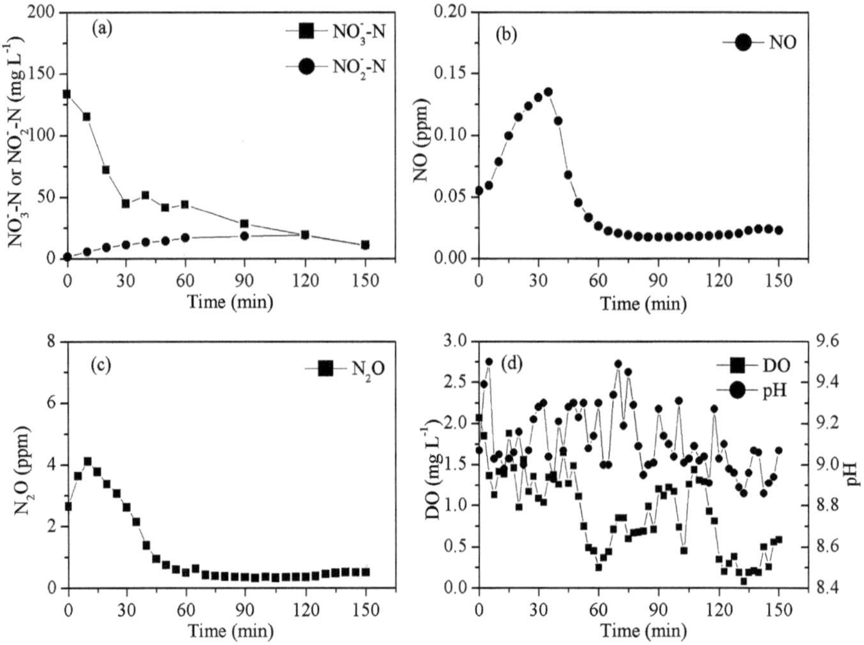

Fig. 4.6 Variations of -N and -N concentration (**a**), NO (**b**) and N_2O (**c**) emission rates, DO and pH (**d**) in R1 under aeration of 10% O_2

4.3.2 The Spatial Distribution of Nitrogen Concentration in the Oxidation Ditch

The COD concentration in effluent was 28.48 ± 5.93 mg L^{-1}, and the removal rate of COD was 88.61% under steady state. The concentrations of ammonia nitrogen, nitrite nitrogen, and nitrite were 0.56 ± 0.37 mg L^{-1}, 1.03 ± 0.02 mg L^{-1}, and 19.45 ± 1.86 mg L^{-1}, respectively. The removal rates of ammonia nitrogen and total nitrogen were 98.89% and 59.92%, showing good nitrification and denitrification function.

The oxidation ditch had the characteristic of a plug flow reactor, so the spatial variation of nitrogen and DO concentrations in the oxidation ditch was measured (Fig. 4.8). Adequate carbon source and anoxic environment in the influent showed that the biological selection pool had a good denitrification function; as a result only 0.01 mg L^{-1} nitrite and 0.59 mg L^{-1} entered into the ditch in the oxidation ditch of the nitrogen (Fig. 4.8a). Excess organic carbon sources were used by bacteria to synthesize PHA as a reserve source of energy and carbon when the bacterial carbon source was insufficient (Fig. 4.9). At the same time, because the oxidation ditch has the characteristics of full-flow reactor, the internal reflux ratio was as high as 62, and the pollutants discharged from the biological selection tank are diluted a lot. Therefore, the concentration of NH_3-N and NO_3-N in region 1 decreased rapidly and

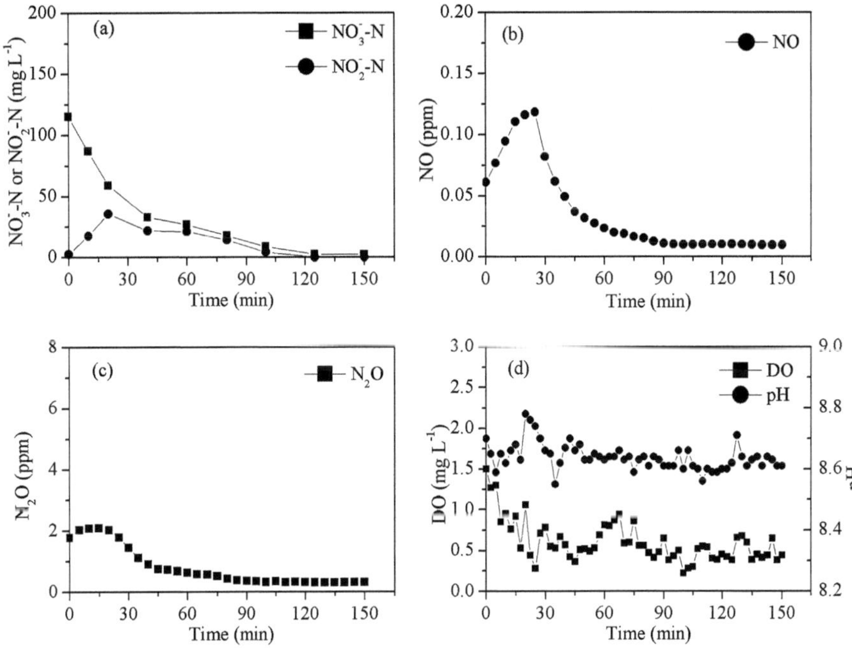

Fig. 4.7 Variations of N concentration (**a**), NO (**b**) and N_2O (**c**) emission rates, DO and pH (**d**) in R2 under aeration of 10% O_2

Table 4.1 N_2O and NO emission in typical cycles of two SBRs under different O_2 concentrations

	N_2O emission(μg)			NO emission(μg)		
O_2 concentration	R1	R2	Reduction %	R1	R2	Reduction %
0%	29.08	23.32	19.81	3.68	3.54	3.80
5%	54.18	29.19	46.12	0.97	1.04	−7.22
10%	14.89	8.17	45.13	0.50	0.44	12.70

increased to 5.06 mg L^{-1} and 12.36 mg L^{-1}, respectively. However, the concentration of nitrite in the region increased to a maximum of 0.16 mg L^{-1} (Hanaki et al. 1992) due to insufficient carbon sources. Dilution effect made the concentration of NH_3-N in the ditch downstream continuously decreased, and the concentration of NO_3-N continued to increase. Once entering the aeration zone, the DO concentration rapidly increased from 0.15 mg L^{-1} to above 0.95 mg L^{-1} (Fig. 4.8b), ensuring normal nitrification in the aeration zone and the constant decline of ammonia nitrogen and nitrite nitrogen concentration in zones 2 and 3. Note that ammonia was completely oxidized in zone 7, whereupon DO rose further, reaching a maximum of 1.77 mg L^{-1} in zone 9. When the mixed solution enters the non-aeration zone, the DO decreased rapidly along the flow direction, and the microorganisms in the sludge used the stored PHA or glycogen for denitrification in the absence of organic carbon source, resulting in the concentration of nitrite (Fig. 4.9).

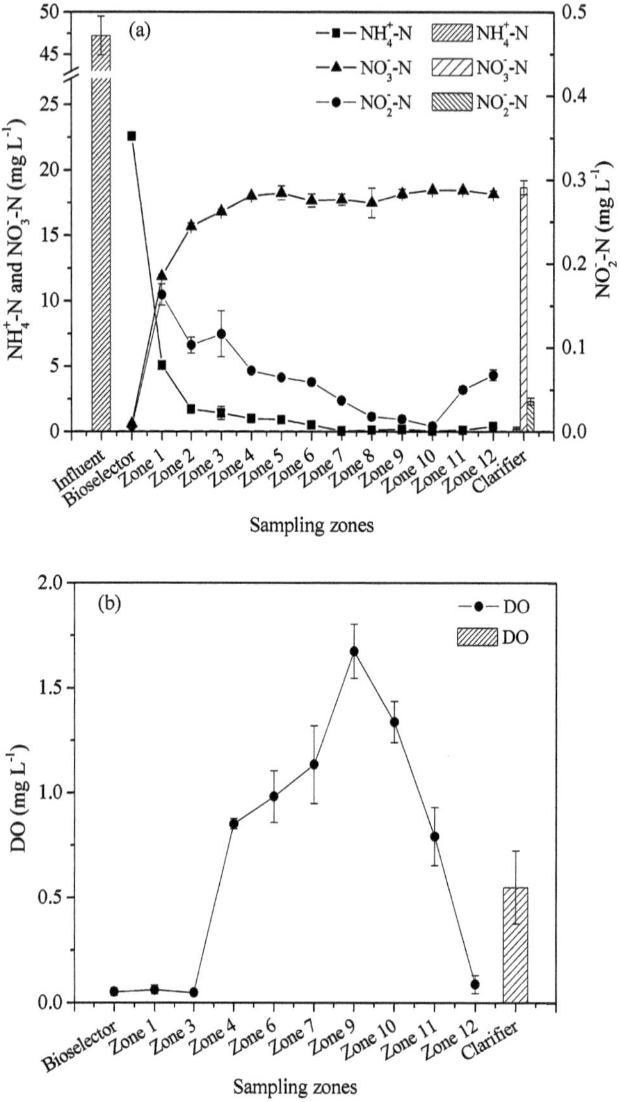

Fig. 4.8 Spatial variations of nitrogen (**a**) and dissolved oxygen (**b**) in the oxidation ditch. (Reprinted from Zheng et al. (2015), with permission from Elsevier)

Consistent with the spatial variability of nitrogen and DO concentrations, N_2O emissions and dissolved N_2O concentrations also exhibited a typical spatial distribution (Fig. 4.10). The N_2O emission rate in the aeration zone (area 4.9) was significantly higher than that in the non-aeration zone ($P < 0.05$), which was mainly due to the strong air-liquid exchange in the aeration zone (Sommer et al. 1998; Lotito et al. 2012). In addition, the follow-up results show that the concentration of DO in

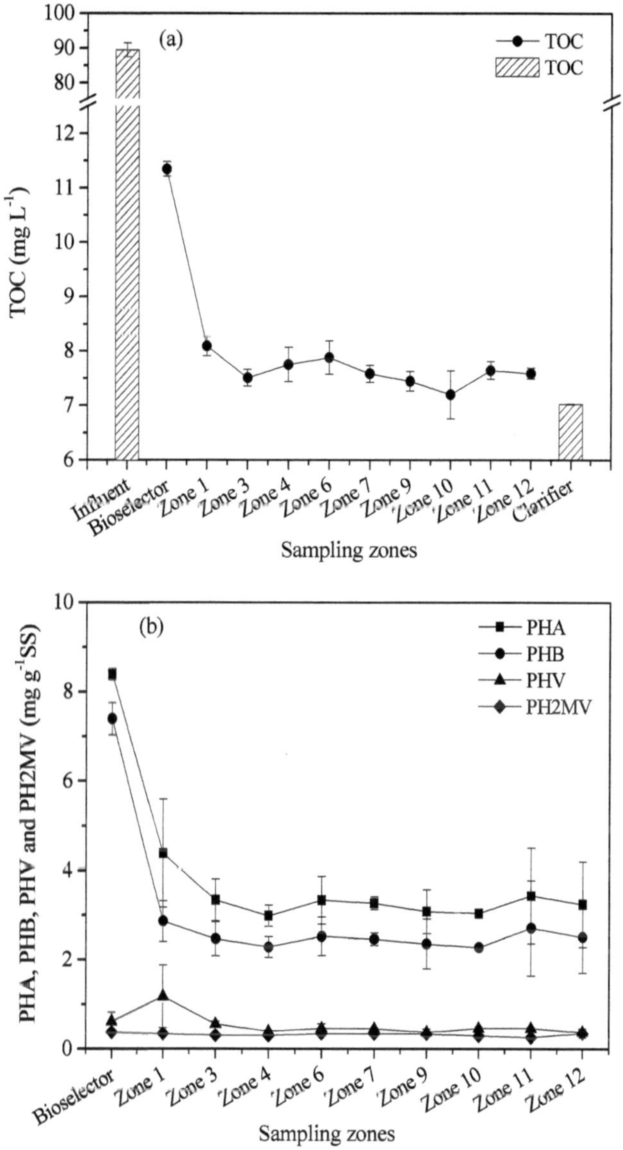

Fig. 4.9 Spatial variations of TOC (**a**) and PHA (**b**) in different zones of the oxidation ditch during steady-state operation. (Reprinted from Zheng et al. (2015), with permission from Elsevier)

the aeration zone was about 1 mg L^{-1}, which is beneficial to N_2O production by autotrophic denitrification. It could be concluded that the aeration zone was not only the main N_2O emission area but also the main N_2O generation area. The N_2O emission rate in the aeration zone decreased significantly along the flow direction ($P < 0.05$), and the N_2O production in the third ditch was significantly lower than

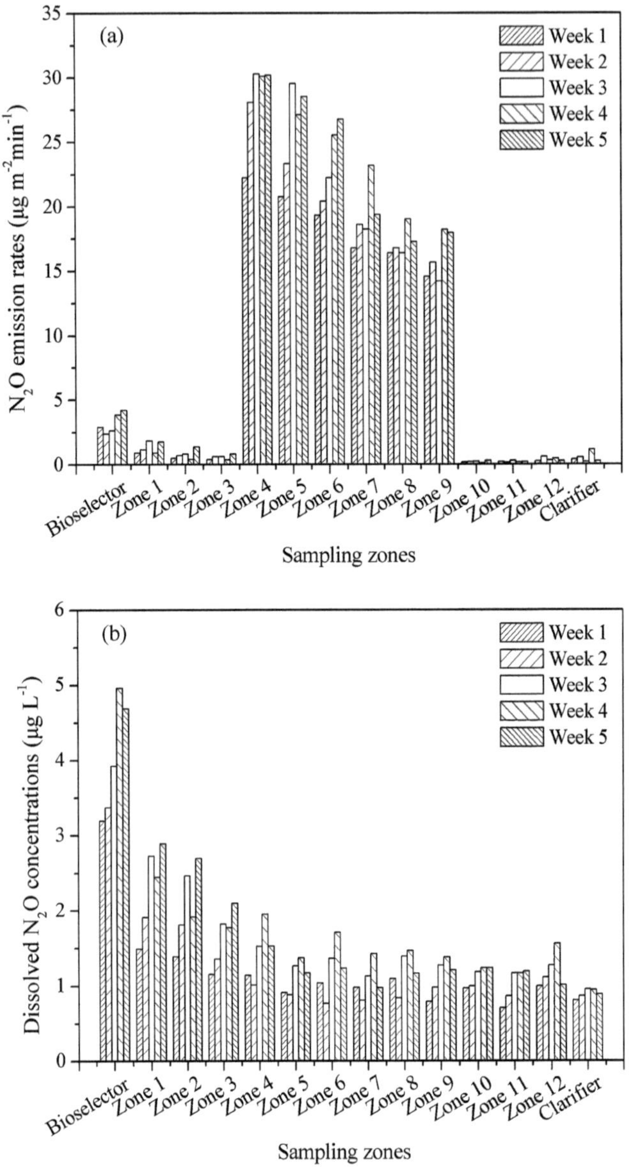

Fig. 4.10 Spatial variations of N_2O emission rates (**a**) and dissolved N_2O concentrations (**b**) in different zones of the oxidation ditch during steady-state operation. (Reprinted from Zheng et al. (2015), with permission from Elsevier)

that in the second one, because the NH_3-N was depleted in the area 7, resulting in an insufficiency in electron donor for N_2O production by autotrophic denitrification. The biological selector is another area with high N_2O emissions, and due to the high

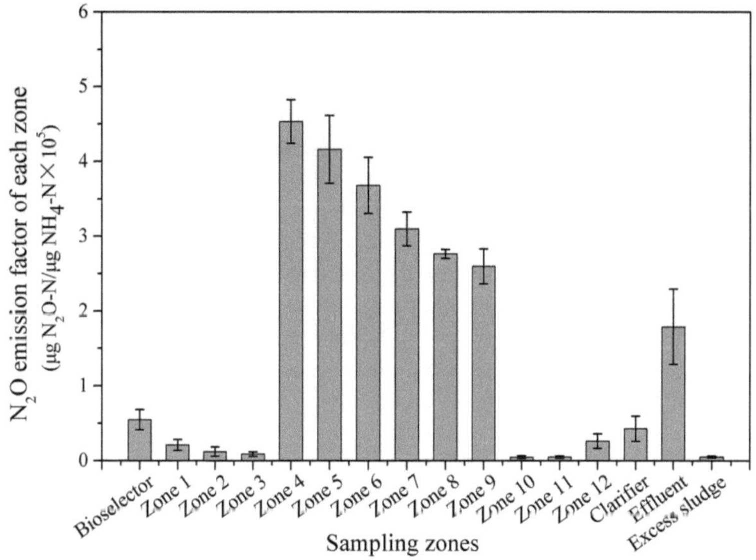

Fig. 4.11 N_2O emission factor of each zone calculated by each emission rate and area. (Reprinted from Zheng et al. (2015), with permission from Elsevier)

dissolved N_2O concentration in the pool (Fig. 4.11), autotrophic denitrification and incomplete heterotrophic denitrification at high carbon loading N_2O emissions were higher than other non-aeration zones (Lotito et al. 2012).

The average gaseous N_2O emission factor for each region was calculated from the emission rate and the area of each sampling region, as shown in Fig. 4.11. The results showed that the contribution of N_2O emissions from the aeration zone to the total N_2O emission from the oxidation ditch was 91.14%. In the non-aeration zone, the contribution of N_2O to total N_2O emissions from the anoxic trench, biological selection tank, and sedimentation tank was 4.59%, 2.40%, and 1.87%, respectively. The total N_2O emission of the whole oxidation ditch system was 0.025% of the total influent nitrogen.

Dissolved N_2O in the mixture was discharged from the system as effluent or excess sludge and eventually released to the ambient atmosphere (Sommer et al. 1998). N_2O was a soluble gas with a saturation solubility of 0.53 mg L^{-1} at 20 °C (Aboobakar et al. 2013). Therefore, the N_2O emissions in the water phase could not be ignored, especially when the sewage treatment capacity was large. In this study, the dissolved N_2O emissions in the effluent account for 0.002% of the total influent nitrogen, accounting for 7.24% of the total N_2O emissions. N_2O emissions from excess sludge only accounted for 0.0001% of the total incoming water, and the remaining sludge production was very small, so this part of the N_2O emissions could be ignored.

It is noteworthy that the total N_2O emission factor of the oxidation ditch system in this study, including both gaseous and liquid forms, was only 0.027%, which was

1–3 orders of magnitude lower than that of other reported wastewater treatment plants 1–2. The recirculation ratio of the oxidation ditch sludge was as high as 62, which made the obvious characteristics of the whole mixed flow and the important reason for the low N_2O emission factor. Nitrogen influent into the oxidation ditch was quickly diluted a lot, making the overall nitrogen concentration in oxidation ditch remains at a low level, such as the ammonia nitrogen, and nitrite concentration in trench was only 5.06 and 0.16 mg L^{-1}, respectively. N_2O produced by autotrophic denitrification or heterotrophic denitrification was very limited (Foley et al. 2010; Lotito et al. 2012) because the concentration of dissolved nitrogen in the water as a substrate was low. In addition, free nitrous acid (FNA) might have an inhibitory effect on the reduction of N_2O (Zhou et al. 2008), and the inhibition of FNA on N_2O reduction under weak alkaline conditions with very low nitrite concentration in the oxidation ditch was also very low.

4.3.3 N_2O Generation Mechanism in Different Areas of the Oxidation Ditch

The denitrification rate and N_2O production rate in different zones of the oxidation ditch were simulated by controlling different dissolved oxygen concentrations in batch experiments (Table 4.2). The results showed that heterotrophic denitrification was only carried out when DO was below 0.2 mg L^{-1} and the rate of heterotrophic denitrification was 0.076 mg-N g^{-1} SS min^{-1} under severe hypoxia conditions. Heterotrophic denitrification resulted in significant N_2O production at the DO concentration of 0.05–0.2 mg L^{-1}, but did not produce N_2O under stringent hypoxia conditions, again confirming the sensitivity of N_2O reductase to O_2 even with weak aerobic. The environment also caused significant accumulation of N_2O during heterotrophic denitrification.

Table 4.2 Nitrogen removal and N_2O generation rates by different biological processes obtained with batch experiments under various DO concentrations

DO	Nitrogen removal rate mg g^{-1}SS min^{-1}			N_2O generation rate µg g^{-1}SS min^{-1}				
mg L^{-1}	HD[a]	NH_3-N oxidation	NO_2-N oxidation	HD[a] + ND[b]	HD[a]	ND[b]	NH_3-N oxidation	NO_2-N oxidation
2.0–3.0	0.000	0.073	0.082	0.142	0.002	0.140	0.131	0.010
1.0–2.0	0.001	0.068	0.072	0.189	0.006	0.183	0.187	0.009
0.5–1.0	0.001	0.046	0.068	0.331	0.001	0.330	0.325	0.005
0.2–0.5	0.000	0.030	0.052	0.309	0.004	0.305	0.126	0.008
0.05–0.2	0.018	0.006	0.034	0.228	0.053	0.175	0.006	0.027
0–0.01	0.076	0.000	0.000	0.010	0.001	0.009	0.000	0.000

[a]HD represented heterotrophic denitrification; [b]ND represented autotrophic denitrification

In the DO range of 0.5–1.0 mg L^{-1}, the N_2O production rate of autotrophic denitrification was the highest, which indicated that the high or low DO was not conducive to autotrophic denitrification. Studies had shown that under certain DO conditions, autotrophic nitrifying bacteria could use nitrite as an electron acceptor, but an overly low DO inhibited the enzyme activity and metabolic processes of bacteria (Kampschreur et al. 2008; Tallec et al. 2006; Yu et al. 2010). The results of this experiment were in agreement with this conclusion. When DO was higher than 0.5 mg L^{-1}, the rate of N_2O produced by autotrophic nitrification was almost equal to the rate of N_2O produced by ammonia oxidation, which was consistent with the results of the inhibition of heterotrophic denitrification in this condition. However, when the DO was lower than 0.5 mg L^{-1}, the rates were very different, which was due to the increase of nitrite concentration by heterotrophic denitrification and the increase of N_2O production. The two biochemical processes together lead to the formation of N_2O.

It is notable that the rate of nitrite oxidation was always higher than that of NH_3-N oxidation. When DO concentration decreased, the difference was larger, which indicated that the process of nitrite oxidation under low DO was more competitive than NH_3-N process. Oxidation ditch sludge in the presence of large amounts of nitrite-oxidizing bacteria (NOB) made nitrite accumulation very small, so the oxidation ditch emissions of N_2O was also very low. In addition, N_2O produced by nitrite oxidation was almost negligible, indicating that ammonia as an electron donor was more important for the production of N_2O by autotrophic denitrification.

The DO distribution in the integrated oxidation ditch and the rate of N_2O production from heterotrophic denitrification and autotrophic denitrification under different DO conditions can be calculated by combining the volume of the mixed solution in each region, and the same was to the amount of N_2O from different sources in each region. The results showed that N_2O production was mainly heterotrophic denitrification in ditch 1 and 12 of the biological selection tank, while in the aeration zone and N4 ditch, the N_2O production was mainly autotrophic denitrification. The contribution of heterotrophic denitrification and autotrophic denitrification to N_2O production was 10.22% and 89.78% for the whole oxidation ditch system, respectively.

4.3.4 *Microbial Community Structure in Oxidation Ditch*

The microbial community structure was analyzed by 16S rRNA high-throughput sequencing (Zhang et al. 2012) to further obtain bio-information related to N_2O emission in oxidation ditch. From the dilution curve, it can be seen that the results provided a comprehensive microbial community structure information (Fig. 4.12) and the information in the level of class and order as shown in Fig. 4.13.

β-Proteobacteria played an important role in the process of nitrogen conversion in sewage treatment including nitrification and denitrification. *Nitrosomonadales* from *β-proteobacteria* was the main genus of AOB genus, accounting for 0.34% of

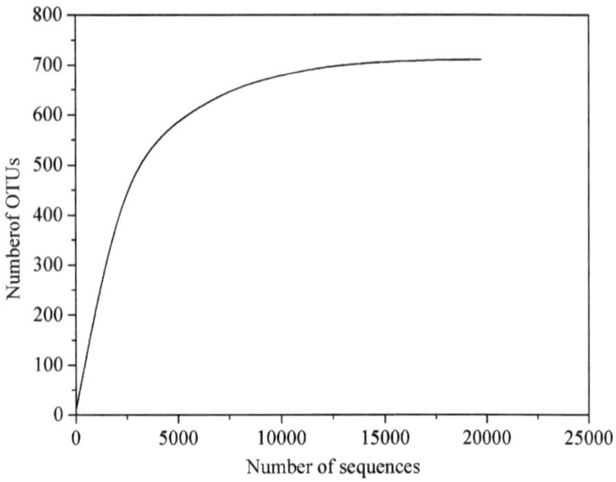

Fig. 4.12 Rarefaction curve of OTUs defined by 3% distances of the activated sludge samples from oxidation ditch. (Reprinted from Zheng et al. (2015), with permission from Elsevier)

the total number of active sludge in oxidation ditch. This part of the bacteria was responsible for the completion of the first step of nitrogen conversion and ammonia nitrogen oxidation for nitrite. In addition, 84.10% of the *β-proteobacteria* species were identified as *Rhodocyclales* (7.90%) and *Burkholderiales* (6.46%). These two highly abundant species contained many bacteria with strong denitrification (Ishii et al. 2009).

In *Rhodocyclales*, *Dechloromonas* (2.57%) and *Zoogloea* (1.39%) were the two genera of the highest abundance, and they were also widely reported in activated sludge. *Zoogloea* played an important role in the flocculation of sludge. *Dechloromonas* has two nirS genes and one atypical nosZ gene, thus ensuring complete denitrification process (Jones et al. 2013). Papirio et al. (2014) detected *Dechloromonas* as a major denitrifying bacterium in a reactor for the treatment of mining wastewater. Based on nirS analysis, Yu et al. (2010) suggested that *Dechloromonas* in the deep subtropical reservoirs dominated denitrifying bacterial community structure. Some researchers have further isolated a strain pure of *Dechloromonas*, and the study found that its denitrification products accounted for the majority of N_2 and therefore had a high N_2O emission reduction potential (Tago et al. 2011).

The *Comamonadaceae* microbes accounted for 78.90% of the *Burkholderiales* in the oxidation ditch samples, which might play an important role in denitrification. *Comamonadaceae* can be enriched with enhanced denitrification carbon source and identified as microorganisms associated with the denitrification process (Ginige et al. 2005). In addition, some other genera of *Burkholderiales* also played an important role in nitrate respiration and N_2O reduction. *Herbaspirillum* enhances denitrification (Ishii et al. 2009) and is the main genus of N_2O-reducing bacteria (Ishii et al. 2011). *Alcaligenes* have excellent heterotrophic nitrification and aerobic denitrification properties and have been used to enhance N_2O emission reduction in biological denitrification systems (Park et al. 2007).

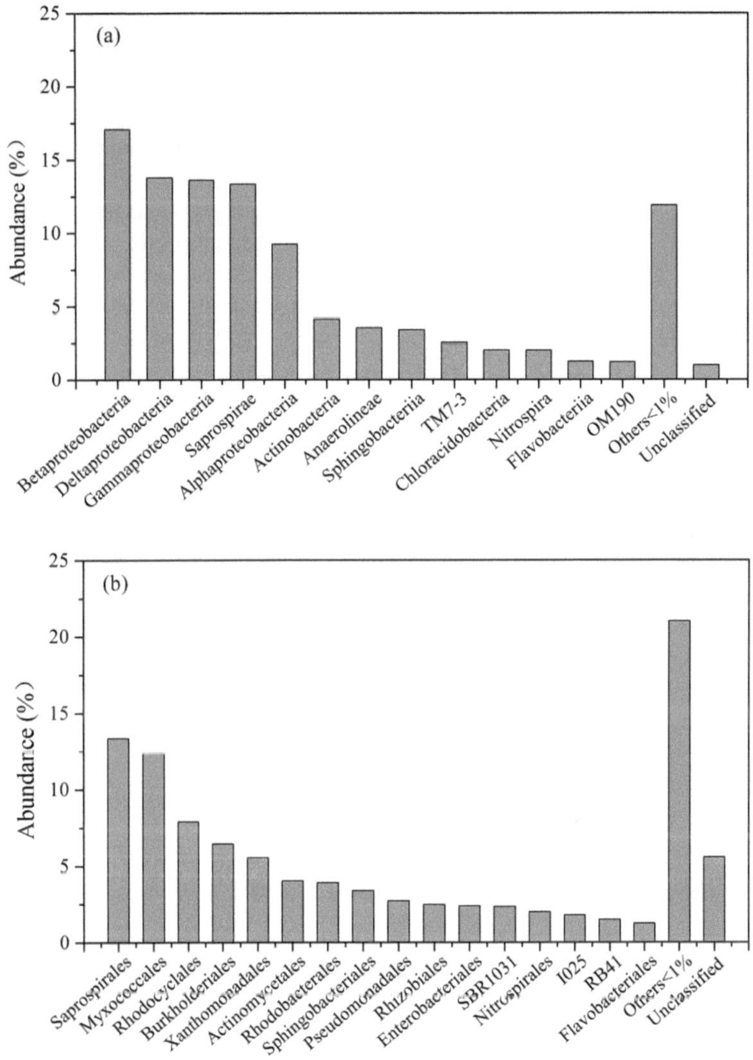

Fig. 4.13 Taxonomic analysis of microbial communities in the oxidation ditch at level of class (**a**) and order (**b**) based on high-throughput 16S rRNA gene sequencing (Reprinted from Zheng et al. (2015), with permission from Elsevier)

Myxococcales accounted for 89.91% of *δ-proteobacteria* and 12.40% of the total sequences. The abundance was relatively high, probably due to the low organic carbon load in the oxidation ditch and moderate DO levels to promote its growth (Zhou et al. 2014). *Myxococcales* was also considered to be one of the major denitrifying bacteria capable of degrading nitrides and nitrous oxides (Manucharova et al. 2000; Zhou et al. 2014), in addition to being generally believed to form mycelia and produce secondary metabolites. *Saprospirale* has the ability to degrade protein and organic particulate matter, and its abundance in the oxidation ditch is also high, probably due to the oxidation ditch-delayed aeration characteristics.

As the most common NOB species, *Nitrospira* in *Nitrospirae* accounted for 1.99% of the total sequence number, which was much higher than that in other sewage plants (Zhang et al. 2012). The reason may be that oxidation ditch operating conditions were very suitable for NOB growth, which greatly enhanced the competitiveness of NOB. First, the concentration of free ammonia and free nitrite in the ditch was very low due to the dilution of the high internal reflux ratio, which had little inhibitory effect on NOB growth (Zhang et al. 2012). Second, there was no rigorous hypoxic environment in the oxidation ditch to inhibit NOB activity (Kornaros et al. 2010), while the presence of nitrite continued to provide a substrate for NOB growth. Finally, a relatively long SRT (25 days) ensured that NOB had sufficient growth time to avoid elution from the system. When the SRT decreased from 25 days to 15 days, the proportion of *Nitrospira* in the total sequence decreased to 0.65.

Therefore, the microbial communities in the oxidation ditch were very rich in this study, and the species of denitrification and N_2O reduction were dominant, which ensured stable nitrogen removal performance and lower N_2O emission. In addition to heterotrophic denitrifying bacteria, autotrophic nitrifying bacteria (NOB), which had a high abundance, also played an important role in reducing NO_2-N and N_2O emissions in the oxidation ditch. Fukumoto et al. (2006) had effectively inhibited the accumulation of nitrite and the emission of N_2O in compost process, using the bio-remediated NOB.

4.3.5 Comparison of N_2O Emission Rate Before and After Bioaugmentation in Oxidation Ditch

It could be seen from the above analysis that the oxidation ditch system was able to achieve good denitrification and control N_2O emission under the condition of 25 days of sludge age. However, in actual activated sludge system, the sludge age was generally selected according to experience, such as AAO process in 8–10 days and oxidation ditch process in 15–25 days, and an overly long sludge age would lead to sludge aging and microbial activity decreasing, especially on the higher requirements of phosphorus removal sewage treatment system. Therefore, sludge age should be shortened. In this study, it was also found that when the oxidation ditch ran at low sludge age (15–20 days), N_2O increased significantly, and the impact load capacity greatly declined. At this point, it was of great significance if the bioremediation could help to achieve N_2O emission reduction and improve the system impact resistance simultaneously.

The immobilized strain PCN-1 was added to the oxidation ditch for bioremediation. The carrier packing could flow in the oxidation ditch with the flow of water, which avoided the loss of the enhanced bacteria without affecting the stability of the

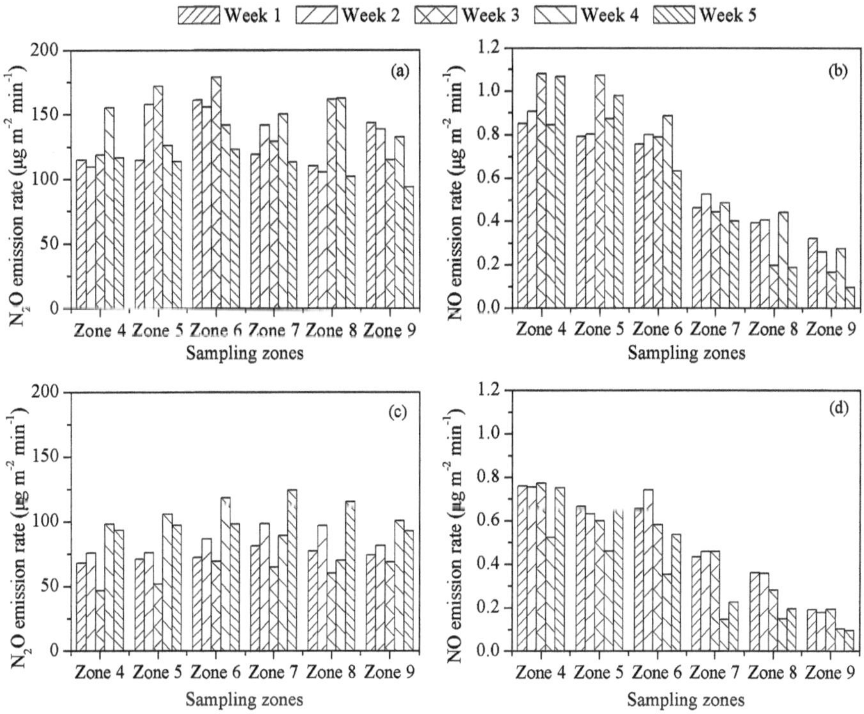

Fig. 4.14 N_2O and NO emission rates from aeration zone of oxidation ditch before (**a** and **b**) and after (**c** and **d**) bioaugmentation with immobilized strain PCN-1

original system. According to the area of the aeration area and the emission rate, the N_2O emission factor in the aeration zone of the oxidation ditch decreased from 0.111% to 0.071%. The total reducing emissions of N_2O and NO were 36.6% and 23.0%, respectively (Fig. 4.14). The results showed that strain PCN-1 could effectively reduce the amount of nitrogen oxides in the oxidation ditch system, which provided a new idea for the reduction of nitrogen oxides in wastewater treatment system.

4.3.6 *Response of Oxidation Ditch Before and After Bioaugmentation to High Ammonia Shock*

In the high ammonia load impact test, the oxidation ditch in the water quality and nitrogen oxide emissions could be divided into three stages (Fig. 4.15): the first stage of 0–2 h, of which the first 1 h for the inflow of high ammonia nitrogen. DO decreased significantly due to the sudden increase of ammonia nitrogen from 1.0 mg L^{-1} to less than 0.1 mg L^{-1} in the first aerobic zone, and the DO of the

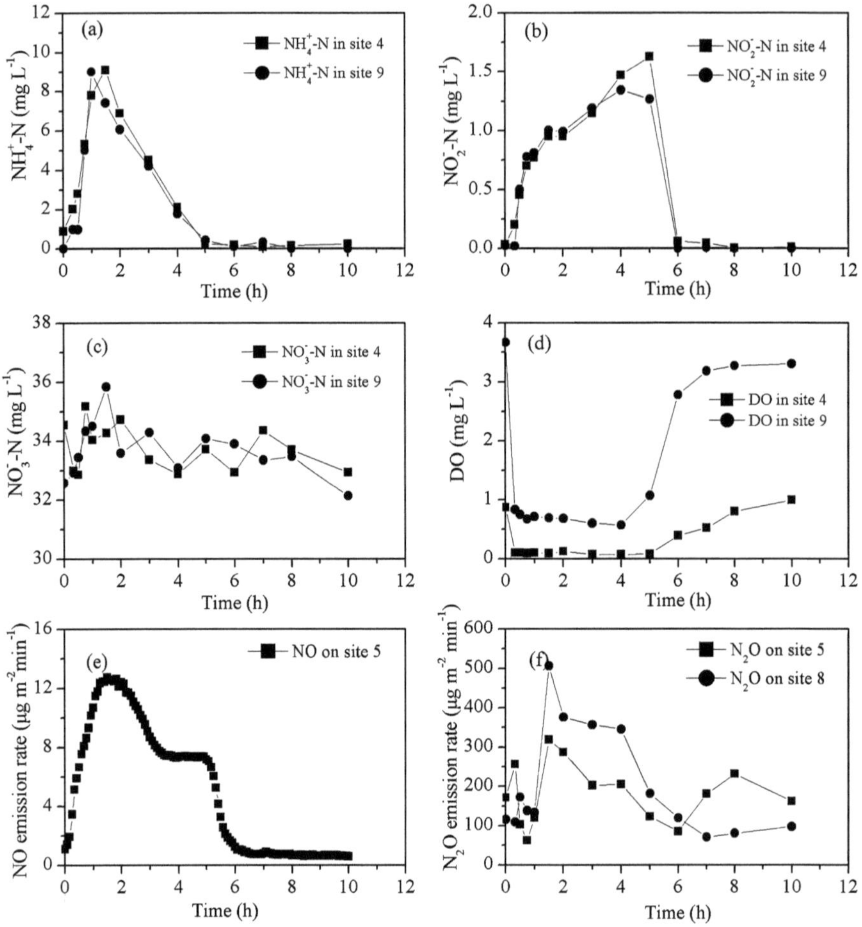

Fig. 4.15 Variations of -N (**a**), -N (**b**), -N (**c**), and DO (**d**); NO (**e**) and N_2O emission rates (**f**) during "ammonia overload shock" experiment before bioaugmentation

aerobic end was also decreased from 3.6 mg L^{-1} to 0.8–0.9 mg L^{-1} and remained at a lower level of less than 1.0 mg L^{-1} for the next 5 h. Ammonia nitrogen concentration in oxidation ditch continued to rise, and the ammonia nitrogen concentration in zone 5 and area 8 rose to the highest value of 9.1 mg L^{-1} during 1.5–2 h. Meanwhile, the concentration of nitrite was greatly increased due to incomplete nitrification and denitrification under low DO concentration. At the same time, NO emission also increased rapidly, and N_2O emission first decreased and then increased rapidly. This was due to the sudden decrease of DO concentration and the increase of NH_3-N and NO_2-N concentration, which enhanced the use of nitrite as an electron acceptor (Kampschreur et al. 2008; Yu et al. 2010). In the second stage, the concentration of NH_3-N decreased rapidly in the two monitoring areas at 2 h and returned to the normal level at 5 h after resuming the normal influent. Although the concentration of nitrite was still increasing, the rate of increase was decreased and began to decrease after 5 h to its peak of 1.6 mg L^{-1}. While the NO and N_2O emission rates showed a decreasing trend, the NO concentration appeared 3–5 h plateau during the descending process, indicating that the oxidation ditch was in the process of recovery. After 5 h, the concentration of NH_3-N in the oxidation ditch was very low, so the rapid increase of DO and the rapid decrease of nitrite concentration occurred. After only 1 h, the DO returned to 3.0 mg L^{-1}. Nitrite concentration decreased to 0.1 mg L^{-1} below. NO and N_2O emission rates also continued to decrease. After 10 h, the oxidation ditch basically restored to the indicators before the impact of the situation.

After the bioaugmentation and stable operation of the oxidation ditch, the high ammonia load impact test was carried out. The indicators showed similar trends with that before bioaugmentation (Fig. 4.16). However, the peak values of ammonia nitrogen and nitrite nitrogen were 6.4 mg L^{-1} and 0.25 mg L^{-1}, which decreased by about 30% and 84%, respectively. This results indicated that the bioaugmentation of strain PCN-1 enhanced the ability of the ditch system to cope with high ammonia load. N_2O emission in the aeration zone also reached the maximum at 1 h, and the peak value was slightly lower than that before bioaugmentation. The rate of N_2O emission decreased rapidly after 1 h and dropped to the level before impact at 2 h, indicating an enhanced recovery ability of the oxidation ditch system under the impact of high ammonia by sudden water flow lift. The total N_2O emission of PCN-1 was reduced by 49.13%, and the total NO emission was reduced by 61.87% after PCN-1 bio-fortification, indicating that the bio-fortification of PCN-1 had led to a significant inhibitory effect on nitrogen oxide increase caused by water influent load impact.

4.3.7 Response of Oxidation Ditch Before and After Bioaugmentation to Aeration Failure

In aeration failure experiment, the DO concentration in the oxidation ditch decreased instantaneously, and the aerobic DO was only 0.3 mg L^{-1}, which led to the weakening of the nitrification ability of the oxidation ditch (Fig. 4.17), but was significantly lower than that at high ammonia loading. For example, the concentrations of ammonia nitrogen and nitrite in zone 4 were 2.1 mg L^{-1} and 0.46 mg L^{-1}, respectively. When the aeration returned to normal after 2 h, ammonia nitrogen and nitrite concentration returned to the level before the impact. The decrease in aeration rate led to a decrease in gas-liquid exchange strength in the oxidation ditch, leading to a reduction in the N_2O blow-off-escape effect. Besides, a low DO concentration (0.1 mg L^{-1} in region 5, 0.5 mg L^{-1} in region 8) reduced the autotrophic denitrification activity of AOB (Yu et al. 2010), which also caused the initial N_2O emission rate decreased. When the aeration returned to normal levels, the DO concentration in zone 8 returned to about 1 mg L^{-1}, which favored the generation of N_2O, and the

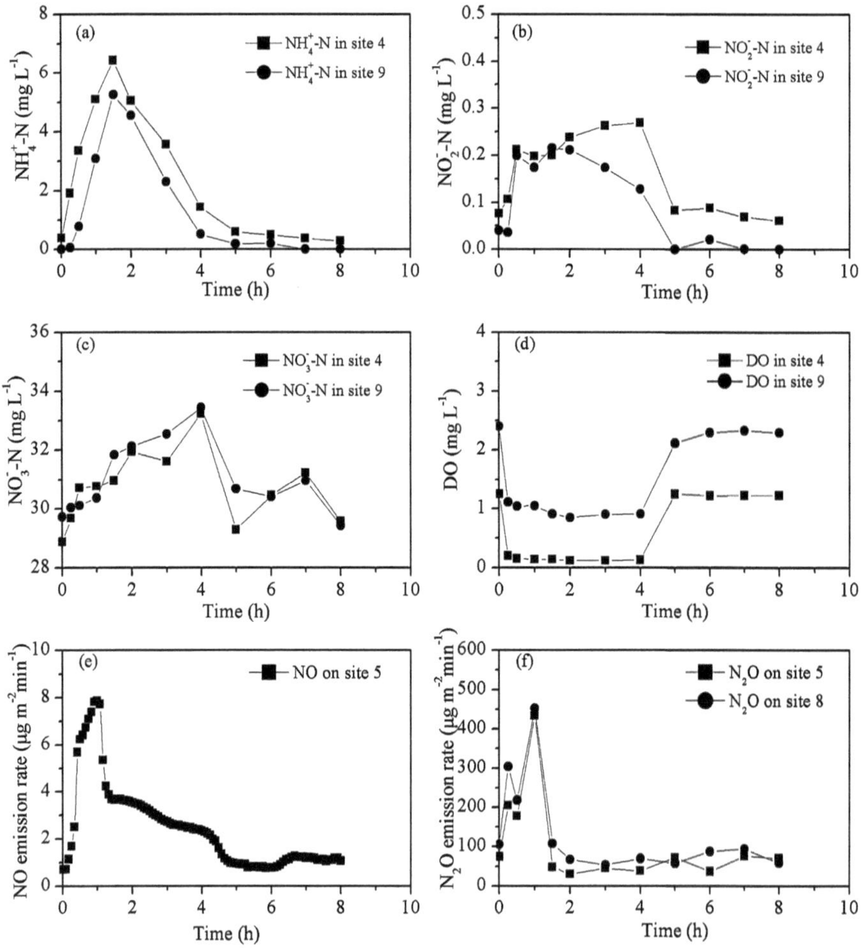

Fig. 4.16 Variations of -N (**a**), -N (**b**), -N (**c**), and DO (**d**); NO (**e**) and N_2O emission rates (**f**) during "ammonia overload shock" experiment after bioaugmentation

N_2O accumulated in the liquid phase was also blown off, resulting in an increase in N_2O emission rate. NO emission rate was increased during the 2 h of aeration reduction and then decreased rapidly to the level before impact after the aeration rate returned to normal.

After the bio-fortification and stable operation of the oxidation ditch, the aeration failure test was carried out again. The trend of various indexes in the oxidation ditch was similar to that before strengthening (Fig. 4.18). DO rapidly decreased, and the accumulation of ammonia nitrogen slightly decreases in nitrite nitrogen and nitrate

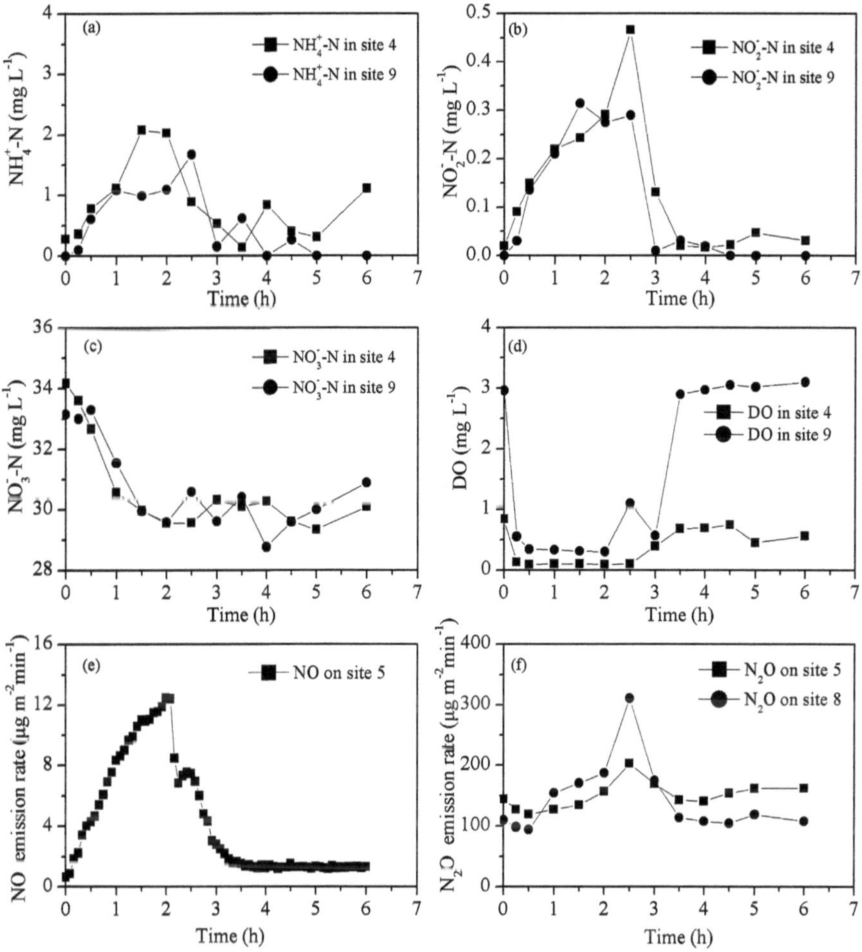

Fig. 4.17 Variations of -N (**a**), -N (**b**), -N (**c**), and DO (**d**); NO (**e**) and N_2O emission rates (**f**) during "aeration failure shock" experiment before bioaugmentation

nitrogen concentration. Compared with that before bioaugmentation, the peak values of ammonia nitrogen and nitrite nitrogen were slightly decreased. NO release remained stable after a rapid rise in 1 h and decreased rapidly to normal after 2 h aeration. The total N_2O emissions from the bio-fortified aeration failure process were reduced by 59.7%, and the total amount of NO was reduced by 64.0%, indicating a significant inhibitory effect of the strain PCN-1 on the massive emission of nitrogen oxides by the aeration failure of the oxidation ditch, enhancing the stability of sewage treatment system is of great significance.

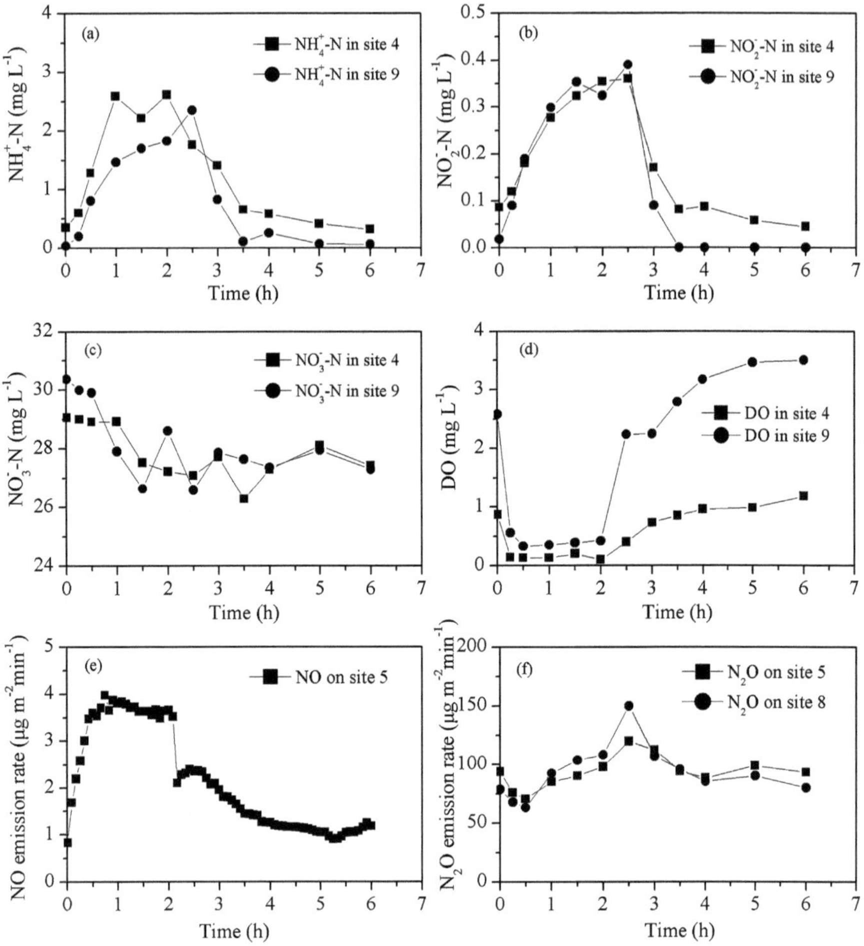

Fig. 4.18 Variations of -N (**a**), -N (**b**), -N (**c**), and DO (**d**); NO (**e**) and N_2O emission rates (**f**) during "aeration failure shock" experiment after bioaugmentation with immobilized PCN-1

4.4 Summary

In this chapter, the application of strain PCN-1 in the bioremediation of activated sludge system was explored. The main conclusions are as follows:

1. The bioremediation of strain PCN-1 could significantly reduce the nitrogen oxide emission during SBR denitrification, and the N_2O emission decreased 19.81%, 46.12%, and 45.13% under the O_2 concentration at 0%, 5%, and 10%, respectively.
2. The Carrousel oxidation ditch has good denitrification effect when its running state is stable. The nitrogen concentration, the DO concentration, the gaseous

N_2O emission rate, and the dissolved N_2O concentration showed a unique spatial distribution. The total N_2O emission of the whole oxidation ditch system accounts for 0.027% of the total influent nitrogen, among which the aeration area accounts for 91.14% and the non-aeration zone such as the hypoxic channel, biological selection tank, and sedimentation tank accounts for 4.59%, 2.40%, and 1.87% respectively.

3. Heterotrophic denitrification had the highest N_2O yield when DO was 0.05–0.2 mg L^{-1} and produced no N_2O under strict hypoxia condition. When DO was 0.5–1.0 mg L^{-1} under autotrophic denitrification, N_2O production rate was the highest, and it was not conducive to autotrophic denitrification with DO concentration too high or too low. The contribution of heterotrophic denitrification and autotrophic denitrification to N_2O production in the whole oxidation ditch system was 10.22% and 89.78%, respectively.
4. The microbial communities in the oxidation ditch were very rich, and the species of denitrification and N_2O reduction are dominant, which ensures the good denitrification performance and N_2O reduction capacity of the system. In addition, the autotrophic nitrifying bacteria NOB of great abundance also played an important role in the oxidation ditch to reduce nitrite concentration and N_2O emissions.
5. When the oxidation ditch system was operated under the steady state of SRT for 17 days, the N_2O emission factor of aeration zone was 0.111%, but the value was decreased to 0.071% after being enhanced by PCN-1. Total emission decreased by 23.0%, showing a good bioaugmentation performance.
6. The oxidation rate of nitrous oxide in the oxidation ditch system was significantly increased, and the N_2O and NO in the oxidation ditch were 49.13% and 69.87%, respectively, after the bioremediation of PCN-1. The N_2O and NO emissions decreased by 59.7% and 64.0%, respectively, under the condition of aeration failure, which showed that the strain PCN-1 played an important role in enhancing the stability of the activated sludge system.

References

Aboobakar, A., Cartmell, E., Stephenson, T., Jones, M., Vale, P., & Dotro, G. (2013). Nitrous oxide emissions and dissolved oxygen profiling in a full-scale nitrifying activated sludge treatment plant. *Water Research, 47*, 524–534.

Beaulieu, J. J., Shuster, W. D., & Febholz, J. A. (2010). Nitrous oxide emissions from a large, impounded river: The Ohio River. *Environmental Science and Technology, 44*, 7527–7533.

Bouchez, T., Patureau, D., Delgenes, J. P., & Moletta, R. (2009). Successful bacterial incorporation into activated sludge floc using alginate. *Bioresource Technology, 100*, 1031–1032.

DeSantis, T. Z., Hugenholtz, P., Larsen, N., Rojas, M., Brodie, E. L., Keller, K., Huber, T., Dalevi, D., Hu, P., & Andersen, G. L. (2006). Greengenes, a chimera-checked 16S rRNA gene database and workbench compatible with ARB. *Applied and Environmental Microbiology, 72*, 5069–5072.

Foley, J., De Haas, D., Yuan, Z., & Lant, P. (2010). Nitrous oxide generation in full scale BNR wastewater treatment plants. *Water Research, 44*, 831–844.

Fukumoto, Y., Suzuki, K., Osada, T., Kuroda, K., Hanajima, D., Yasuda, T., & Haga, K. (2006). Reduction of nitrous oxide emission from pig manure composting by addition of nitrite-oxidizing bacteria. *Environmental Science and Technology, 40*, 6787–6791.

Ginige, M. P., Keller, J., & Blackall, L. L. (2005). Investigation of an acetate-fed denitrifying microbial community by stable isotope probing, full-cycle rRNA analysis, and FISH. *Applied and Environmental Microbiology, 71*, 8683–8691.

Hanaki, K., Hong, Z., & Mstsuo, T. (1992). Production of nitrous oxide gas during denitrification of wastewater. *Water Science and Technology, 26*, 1027–1036.

Ishii, S., Yamamoto, M., Kikuchi, M., Oshima, K., Hattori, M., Otsuka, S., & Senoo, K. (2009). Microbial populations responsive to denitrification-inducing conditions in rice paddy soil, as revealed by comparative 16S rRNA gene analysis. *Applied and Environmental Microbiology, 75*, 7070–7078.

Ishii, S., Ohno, H., Tsuboi, M., Otsuka, S., & Senoo, K. (2011). Identification and isolation of active N_2O reducers in rice paddy soil. *ISME Journal, 5*, 1936–1945.

Jones, C. M., Graf, D. R. H., Bru, D., Philippot, L., & Hallin, S. (2013). The unaccounted yet abundant nitrous oxide-reducing microbial community: A potential nitrous oxide sink. *The ISME Journal, 7*, 417–426.

Kampschreur, M. J., Tan, N. C., Kleerebezem, R., Picioreanu, C., Jetten, M. S. M., & Van Loosdrecht, M. C. M. (2008). Effect of dynamic process conditions on nitrogen oxide emission from a nitrifying culture. *Environmental Science and Technology, 42*, 429–435.

Kornaros, M., Dokianakis, S. N., & Lyberatos, G. (2010). Partial nitrification/denitrification can be attributed to the slow response of nitrite oxidizing bacteria to periodic anoxic disturbances. *Environmental Science & Technology, 44*, 7245–7253.

Lotito, A. M., Wunderlin, P., Joss, A., Kipf, M., & Siegrist, H. (2012). Nitrous oxide emissions from the oxidation tank of a pilot activated sludge plant. *Water Research, 46*, 3563–3573.

Manucharova, N. A., Dobrovol'skaya, T. G., & Stepanov, A. L. (2000). Taxonomic composition of denitrifying bacteria in soddy podzolic soil. *Microbiology, 69*, 234–237.

Papirio, S., Ylinen, A., Zou, G., Peltol, M., Esposito, G., & Puhakka, J. A. (2014). Fluidizedbed denitrification for mine waters. Part I: Low pH and temperature operation. *Biodegradation, 25*, 425–435.

Park, K. Y., Kim, S. J., Jung, J. Y., & Lee, S. H. (2007). Reduction of N_2O emission from biological nitrogen removal processes by Alcaligenes faecalis augmentation. *Journal of Industrial and Engineering Chemistry, 13*, 508–511.

Sommer, J., Ciplak, A., Sumer, E., Benckiser, G., & Ottow, J. C. G. (1998). Quantification of emitted and retained N_2O in a municipal wastewater treatment plant with activated sludge and nitrification-denitrification units. *Agrobiological Research, 51*, 59–73.

Tallec, G., Garnier, J., Billen, G., & Gousailles, M. (2006). Nitrous oxide emissions from secondary activated sludge in nitrifying conditions of urban wastewater treatment plants: effect of oxygenation level. *Water Research, 40*, 2972–2980.

Tago, K., Ishii, S., Nishizawa, T., Otsuka, S., & Senoo, K. (2011). Phylogenetic and functional diversity of denitrifying bacteria isolated from various rice paddy and rice-soybean rotation fields. *Microb Environ, 26*, 30–35.

Weiss, R. F., & Price, B. A. (1984). Nitrous oxide solubility in water and seawater. *Marine Chemistry, 8*, 347–359.

Wunderlin, P., Mohn, J., Joss, A., Emmenegger, L., & Siegrist, H. (2012). Mechanisms of N_2O production in biological wastewater treatment under nitrifying and denitrifying conditions. *Water Research, 46*, 1027–1037.

Yang, Q., Liu, X., Peng, C., Wang, S., Sun, H., & Peng, Y. (2009). N_2O production during nitrogen removal via nitrite from domestic wastewater: main sources and control method. *Environmental Science and Technology, 43*, 9400–9406.

Yang, X. P., Wang, S. M., Zhang, D. W., & Zhou, L. X. (2011). Isolation and nitrogen removal characteristics of an aerobic heterotrophic nitrifying-denitrifying bacterium, *Bacillus subtilis* A1. *Bioresource Technology, 102*, 854–862.

Yu, R., Kampschreur, M. J., Loosdrecht, M. C. M. V., & Chandran, K. (2010). Mechanisms and specific directionality of autotrophic nitrous oxide and nitric oxide generation during transient anoxia. *Environmental Science and Technology, 44*, 1313–1319.

Zhang, T., Shao, M. F., & Ye, L. (2012). 454 Pyrosequencing reveals bacterial diversity of activated sludge from 14 sewage treatment plants. *ISME Journal, 6*, 1137–1147.

Zheng, M. S., Tian, Y. H., Liu, T., Ma, T., Li, L., Li, C., Ahmad, M., Chen, Q., & Ni J.R.*. (2015). Minimization of nitrous oxide emission in a pilot-scale oxidation ditch: Generation, spatial variation and microbial interpretation. *Bioresource Technology, 179*, 510–517.

Zhou, Y., Pijuan, M., Zeng, R. J., & Yuan, Z. G. (2008). Free nitrous acid inhibition on nitrous oxide reduction by a denitrifying-enhanced biological phosphorus removal sludge. *Environmental Science and Technology, 42*, 8260–8265.

Zhou, X. W., Li, S. G., Li, W., Jiang, D. M., Han, K., Wu, Z. H., & Li, Y. Z. (2014). Myxobacterial community is a predominant and highly diverse bacterial group in soil niches. *Environmental Microbiology Reports, 6*, 45–56.

Chapter 5
Denitrification Characteristics and Applications of *Pseudomonas aeruginosa* PCN-2

5.1 Introduction

NO and NO_2 (NO_x) in the atmosphere are major pollutants which cause serious environmental problems such as acid deposition, photochemical smog, and ozone hole. Fossil fuel combustion is the main source of NO_x. Biological technology is an alternative for NO_x removal from flue gas due to its low cost, high efficiency, and easy operation (Barnes et al. 1995). Biofilter and biotrickling filter are the two most commonly used structures, in both of which the denitrifying bacteria are immobilized onto the packing material in the reactor and reduce the NO_x to N_2 (Jin et al. 2005). However, the flue gas usually contains 2–10% oxygen, which would inhibit the denitrification process and result in poor NO removal efficiency (Lee and Apel 1998; Niu et al. 2014). Herein, the unique advantages of aerobic denitrification provide a new way to solve this problem (Jiang et al. 2009).

In this chapter, the aerobic nitrogen removal characteristics and mechanisms of *Pseudomonas aeruginosa* PCN-2 were explored systematically by batch experiments and gene expression analysis, where the effect of oxygen on denitrification and NO_x removal was emphasized. A biotrickling filter system for flue gas purification was constructed with immobilized strain PCN-2, and the NO_x removal efficiency under different inlet NO and O_2 concentrations was investigated. This study provided an experimental and theoretical basis for highly efficient flue gas purification by aerobic denitrification.

M. Zheng, *Nitrogen Removal Characteristics of Aerobic Denitrifying Bacteria and Their Applications in Nitrogen Oxides Emission Mitigation*, Springer Theses,
https://doi.org/10.1007/978-981-13-2432-1_5

5.2 Materials and Methods

5.2.1 Media

The composition of medium used for bacterial cultivation and aerobic denitrification evaluation (per liter) was as follows: 13.51 g $C_4H_4Na_2O_4$, 0.63 g NH_4Cl, 0.61 g $NaNO_3$ or 0.39 g $NaNO_2$, 1.76 g $K_2HPO_4{\cdot}3H_2O$, 0.50 g $MgSO_4{\cdot}7H_2O$, 0.02 g $CaCl_2$, 0.005 g $FeSO_4{\cdot}7H_2O$, and 0.1 mL trace elements (same composition as 2.2.2).

The composition of medium used for biotrickling filter nutrient solution was as follows (per liter): 1.2 g $C_6H_{12}O_6$, 0.06 g $NaNO_3$, 0.19 g NH_4Cl, 1.0 g $K_2HPO_4{\cdot}3H_2O$, 0.20 g $MgSO_4{\cdot}7H_2O$, 0.02 g $CaCl_2$, 0.005 g $FeSO_4{\cdot}7H_2O$, and 0.1 mL trace elements.

The pH of all media was adjusted to 7.5 and sterilized at 121 °C for 30 min before use.

5.2.2 Aerobic Denitrification Performance of Strain PCN-2

In order to study the aerobic denitrification performance of PCN-2 and quantify the gas product accurately, the experiment was carried out in 300 mL sealed serum bottles. The bottle was filled with 80 mL DM, and after replacing the air of the upper space with high-purity helium, 50 mL of pure oxygen was injected into the bottle. The oxygen in the upper part of the serum bottle was about 21%, similar with the oxygen concentration in air. The injection of oxygen to the serum created a positive pressure state, thus avoiding the interference of nitrogen from external air and ensure the accurate quantitative of N_2. In order to study the denitrification performance of PCN-2 under different oxygen concentrations, the O_2 content in the upper space of the serum bottle was set as 0%, 10%, 20%, 50%, and 100% by injecting different volume of pure oxygen.

An 8 mL bacterial suspension at late exponential growth phase was collected, washed twice with DD H_2O, resuspended, and then inoculated into a serum bottle. The bottles were incubated in a rotary shaker at 150 rpm and 30 °C. At regular intervals, 250 μL gas samples were collected from the serum bottle with a gas-tight syringe for measuring N_2O, O_2, and N_2 concentrations, and 2 mL gas samples was withdrawn for measurement of NO concentration. Simultaneously, 2 mL aliquots were withdrawn and centrifuged at 8000 rpm for 5 min. The supernatant was used to measure the concentration of ammonia nitrogen, nitrite, and nitrate. The remaining pellets were resuspended in RNA store solution, placed at 4 °C overnight, and stored at −20 °C for RNA extraction.

In order to study the application of the strain in wastewater treatment, 7.5 mL and 15 mL bacterial suspension were collected after being washed and then resuspended

Table 5.1 PCR primers of 16S rRNA and denitrifying genes for strain PCN-2

Gene name	Primer name	Primer sequence (5′-3′)	Amplification length (bp)	References
16S V3	F341	CCTACGGGAGGCAGCAG	177	Muyzer et al. (1993)
	R518	ATTACCGCGGCTGCTGG		
napA	napA V17F[a]	TGGACCATGGGCTTCAACC	152	Bru et al. (2007)
	napA 4R[a]	ACCTCGCGCGCGGTGCCGCA		
nirS	nirS cd3aF[a]	GTCAACGTGAAGGAGACCGG	425	Throback et al. (2004)
	nirS R3cd[a]	GACTTCGGATGGGTCTTGA		
cnorB	cnorB 2F[a]	GACAAGTTCTACTGGTGGT	146	Braker and Tiedje (2003)
	cnorB 7,1R[b]	GCGATGATCACGTAGAGCCA		
nosZ	nosZ 1527F	CGCTGTTCHTCGACAGYCA	246	Throback et al. (2004)
	nosZ 1773R	ATRTCGATCARCTGBTCGTT		

[a]Designed by *nirS* gene sequence of *P. aeruginosa* PAO1
[b]Designed by primer designing tool in NCBI according to *cnorB* sequence of *P. stutzeri* A1501

in a 300 mL serum bottle containing 50 mL activated sludge (MLVSS = 1208 mg L^{-1}). The inoculation dosages (v/v%) were 15% and 30%, and the strain concentrations were 113 mg VSS L^{-1} and 226 mg VSS L^{-1}, respectively (as calculated by Fig. S2). At the same time, the control group was supplemented with the appropriate amount of activated sludge that equalizes the total bacterial amount.

5.2.3 Real-Time Quantitative PCR Analysis

Total RNA extraction and cDNA synthesis were synthesized using RNAprep Bacteria Kit and FastQuant RT Kit (TianGen, China), respectively. Real-time PCR was performed using SYBR Green PCR Master Mix (Applied Biosystems) and designed *napA*, *nirS*, *cnorB*, and *nosZ* primers (Table 5.1). V3 region of 16S rRNA was used as the internal standard to normalize for differences of cDNA added to each PCR tubes. Reactions were run in a real-time PCR system (Applied Biosystems 7300) under the following conditions: initial denaturation at 95 °C for 10 min, followed by 40 cycles of denaturation at 95 °C for 10 s, annealing at 59 °C (V3, *nirS*, and *nosZ*) or 63.5 °C (*napA*) or 56 °C (*cnorB*) for 20 s, and extension at 72 °C for 31 s.

5.2.4 Sealed Serum Bottle Experiments of Immobilized Strain PCN-2

NO_x removal performance by immobilized strain PCN-2 was firstly investigated in batch experiments with fixed sealed bottles. The functional polyurethane foam was chosen as bacterial immobilization carrier due to the advantages of high-mass transfer rate and pollutant removal efficiency ascribed to its large surface area and high bacteria-loading capacity (Cheng et al. 2010; Woertz et al. 2002). Sterilized cubic foams with sides of approximately 1 cm were put into the basal medium in 500 ml flasks inoculated with strain PCN-2. The foams were incubated for biofilm growth for 2 days and then transferred to duplicate 500 ml sealed bottles to 10% of the bottle volumes. The gas in the bottles were replaced by helium gas and injected into different volume of pure O_2 to reach O_2 proportion of 0, 2.5, 5.0, 7.5, and 10.0%, respectively. The bottles with carriers soaking in medium without bacterial inoculation were set as control.

At time 0 min, pure NO gas was injected into the bottles to reach the concentration of 3000 ppm. After 1 h, 250 μL gaseous samples were collected for N_2O and N_2 analysis and 2 mL for determining the concentrations of the remaining NO and NO_2. Then NO gas was injected again to reach around 3000 ppm and analyzed after 1 h; the same procedure was repeated four times to test the bacterial capacity of NO removal. Finally, the liquid in the foam was squeezed out to measure NH_4^+-N, NO_2^--N, and NO_3^--N concentrations.

5.2.5 Application of Immobilized Strain PCN-2 in Trickling Filter

The NO_x removal in simulated flue gas of strain PCN-2 was investigated by applying it in a biotrickling filter. A biotrickling filter with 40 cm height, 5.0 cm inner diameter, and 30 cm bed depth was made by plexiglass and filled the functional polyurethane foam as packing material. Simulated flue gas was provided by the NO gas and ambient air in different proportions and pumped to the bottom of the filter at a total rate of 350 mL min^{-1}, and off-gas was released to the atmosphere through the outlet at the top. The trickling media stored in a 5 L sealed bottle which had been inoculated by strain PCN-2 were pumped into the top of the reactor, sprinkled on the packed bed, and then gravitationally flowed back to the media storage bottle. A 500 mL medium was replaced by a new sterilized medium every day to ensure efficient carbon supply and toxic metabolite discharge.

The reactor had been operated for 2 weeks to achieve steady state before starting the following experiments. NO concentrations at 100, 200, 300, 400, and 500 ppm were employed to investigate NO removal capacity of the biotrickling filter under inlet O_2 concentration of 5%. In the O_2 influence experiments, O_2 proportion at 0%, 2.5%, 5.0%, 7.5%, and 10.0% in the inflow were tested under inlet NO concentration of 300 ppm. The off-gas was collected from the outlet for measuring NO_x concentrations every day, and the system was operated for at least 5 days under each configuration to achieve steady state.

5.2.6 Analytical Methods

The analytical and calculation methods of OD_{600}, dissolved oxygen (DO), NH_4^+-N, NO_2^--N, NO_3^--N, O_2, N_2O, N_2, and NO_x were the same as 3.2.5.

5.3 Results and Discussion

5.3.1 Aerobic Nitrate Removal by Strain PCN-2

It could be seen that strain PCN-2 removed all of the NO_3-N in the 6 h after inoculation and reduced the oxygen concentration from 17.35% to 9.62% (Fig. 5.1a), indicating that the strain PCN-2 used oxygen and nitrogen oxide simultaneously as the electron acceptor for growth and metabolism which means aerobic denitrification was performed (Robertson et al. 1995). With the removal of NO_3-N, nitrite gradually accumulated and reached 28.74 mg L^{-1} at 4 h and then began to decrease and completely deplete at 8 h (Fig. 5.1b).

In addition to monitoring nitrogen transformation process in medium, the accurate quantitation of gaseous intermediates NO, N_2O, and N_2 in the denitrification was also carried out for the entire characteristics of aerobic denitrification. Because the strain did not have the function of heterotrophic nitrification, NO gaseous products, N_2O and N_2, could only be generated from the degradation of nitrate nitrogen, and the assimilation of ammonia nitrogen was used for microbial growth (Knowles 1982). As shown in Fig. 5.1b, N_2O kept accumulating along with the degradation of nitrate nitrogen and nitrate nitrogen with the highest concentration of 81.08 mg in 8 h, L^{-1}; most of the initial nitrate nitrogen (100 mg L^{-1}) had been transformed to N_2O. Since then, the N_2O concentration has been slowly declining and started to be consumed rapidly when the oxygen concentration declined below 2.0% at 21 h. The final concentration of N_2 was 108.17 mg L^{-1}, which was consistent with the initial nitrate concentration.

Interestingly, as the characteristics of denitrification processed intermediately, no NO in strain PCN-2 during the entire process of denitrification were detected, resulting in an obvious difference from the majority of aerobic denitrifying bacteria. For example, NO generated by the strain *Alcaligenes eutrophus* LMD 82.41 accounted for 0.39% of the total nitrogen (Kester et al. 1997). The max accumulation of NO generated by *P. Stutzeri* PCN-1 in this study was 3.33 μg L^{-1} during the process of nitric nitrogen degradation. NO is a highly reactive, free, diffused gas molecule that could produce significant toxic effects within the cell. It was reported

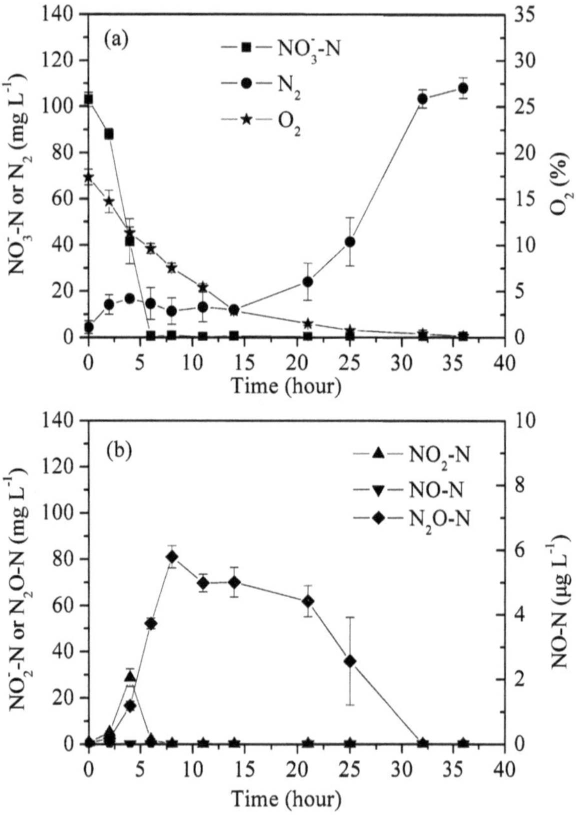

Fig. 5.1 Nitrogen transformations during aerobic nitrate removal by strain PCN-2. (Reprinted from Zheng et al. (2016), with permission from Elsevier)

that the NO at 0.05 mmol L^{-1} could completely inhibit the transcription of *nirS* and *nor* of *Pseudomonas stutzeri* (Vollack and Zumft 2001). It was possible to infer that the great NO reduction capability of strain PCN-2 may be due to the synergistic expression and regulatory action of the nirS and norB genes (Spiro 2012).

5.3.2 Aerobic Nitrite Removal by Strain PCN-2

The aerobic denitrification properties of the strain PCN-2 when using nitrite as nitrogen sources were investigated (Fig. 5.2). Nitrite in medium reduced quickly after inoculation and was completely removed within 6 h; corresponding to the removal rate was 14.65 mg L^{-1} h^{-1} much higher than that of other aerobic denitrification strains. For example, *Pseudomonas* yy7 had a degradation rate of only 0.76 mg L^{-1} d^{-1} when nitrite was the nitrogen source (Wan et al. 2011). The process of denitrification where nitrite was the nitrogen source was similar to that of nitrate nitrogen as nitrogen source. Although NO accumulation has been detected in this process, it was only at 2 h when it reached to 2 μg L^{-1} and might be derived from

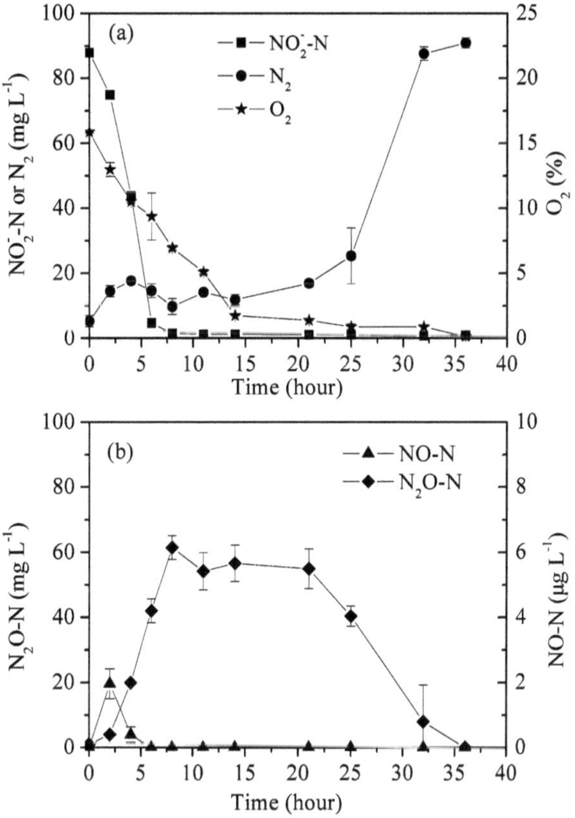

Fig. 5.2 Nitrogen transformations during aerobic nitrite removal by strain PCN-2. (Reprinted from Zheng et al. (2016), with permission from Elsevier)

the dismutation of the free HNO_2. In general, NO reductase of aerobic denitrifying bacteria would be seriously inhibited by nitrite and oxygen. Bergaust et al. (2008) found that the root carcinoma agrobacterium C58 in 7% oxygen and nitrite nitrogen concentration of 2 mmol L^{-1} produced 40 μmol L^{-1} NO, accounting for 78.26% of the total nitrate nitrogen. It was tempting to speculate that the aerobic respiration and denitrifying enzyme activity of strain PCN-2 would not be impacted by the high concentration of nitrite, which enhanced the strain's capability to survive application potential in various environments.

5.3.3 *Expressions of Denitrification Genes of Strain PCN-2*

The quantitative analysis of the four denitrifying genes expressions by strain PCN-2 during nitrate reduction were displayed (Fig. 5.3). The results of the agar gel electrophoresis showed that four functional genes were expressed successfully in the denitrification process (Fig. 5.4). It could be observed that gene *napA*, *nirS*, and

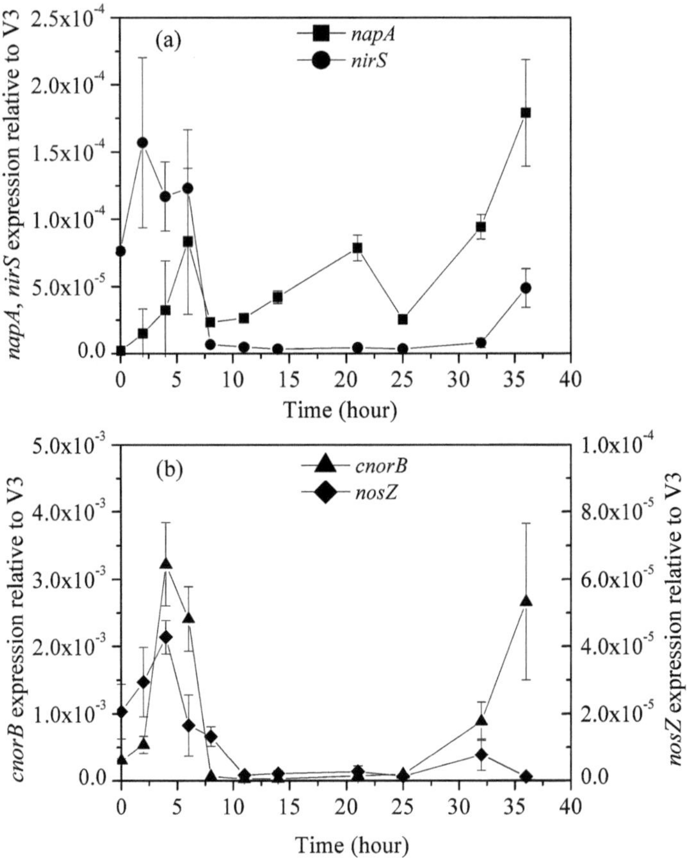

Fig. 5.3 Expressions of denitrification genes during aerobic nitrate reduction by strain PCN-2. (Reprinted from Zheng et al. (2016), with permission from Elsevier)

cnorB expressions were on the rise during the first 6 h after inoculation, which was consistent with the gradual reduction of the nitrate nitrogen, NO and N_2O (Fig. 5.1), and it could be concluded that the three kinds of gene expression and coordinated control ensured the smooth transformation of the nitrogen. With the consumption of nitrate nitrogen, nitrite nitrogen, and other denitrifying substrate, mRNA was unable to be synthesized due to the lack of appropriate inducing substrate, and the expression of three kinds of functional genes also began to decrease rapidly (Philippot 2002). However, in the last 2 h, three kinds of gene expression showed significant rise, and this might be due to the substrate consumption to bacteria in order to maintain normal metabolism of life and more functional gene transcription (Lu et al. 2014). This conclusion still needs to be further confirmed.

It is worth noting that gene expression of cnorB was one order of magnitude higher than nirS gene, which showed that strain PCN-2 might synthesize a lot of NO reductase and prevent the accumulation of NO (Spiro 2012). Compared with the

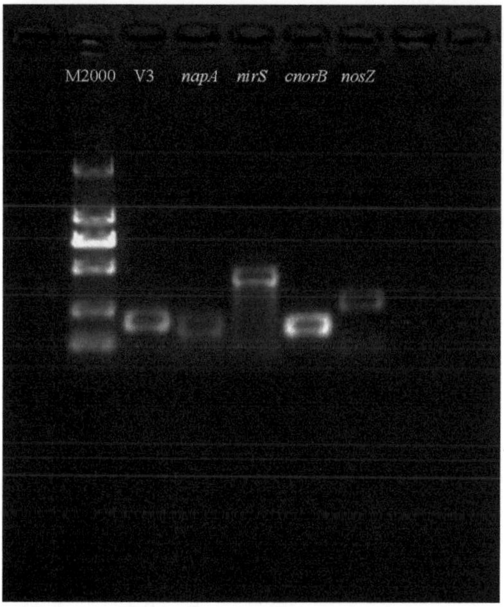

Fig. 5.4 Agarose gel electrophoresis of real-time quantitative PCR amplification of denitrifying genes. (Reprinted from Zheng et al. (2016), with permission from Elsevier)

previous research results, the bacteria also showed obvious advantages of aerobic denitrification at the genetic level. For example, the denitrification genes of *P. fluorescens* C7R12 could not express under the aerobic conditions (Philippot 2002) neither could cnorB genes of strain of *P. mandelii* (Saleh-Lakha et al. 2008). Although *nosZ* gene in the reduction process of nitrate nitrogen also presented a similar trend, the quantity was one or two orders of magnitude lower when compared with the above three gene expression, which was probably ascribed to oxygen inhibition. The conversion of N_2O to N_2 was hindered as well. After complete consumption of nitrate nitrogen and nitrite nitrogen, the expression quantity of *nosZ* gene had been maintained at a very low level. The relatively high gene expression at 32 h was possibly due to the massive oxygen consumption and high N_2O substrate concentration, which was the main reason for the conversation of N_2O to N_2 at this time.

5.3.4 Denitrification Performances of Strain PCN-2 Under Different Oxygen Concentrations

Strain PCN-2 showed significant denitrification difference under different oxygen concentrations (Fig. 5.5). Under all the oxygen concentration gradient, nitrate nitrogen concentrations decreased rapidly after inoculation, while the initial nitrate consumption rate had decreased with the rise of oxygen concentration. However, nitrate nitrogen could also be removed completely in 9 h after vaccination even

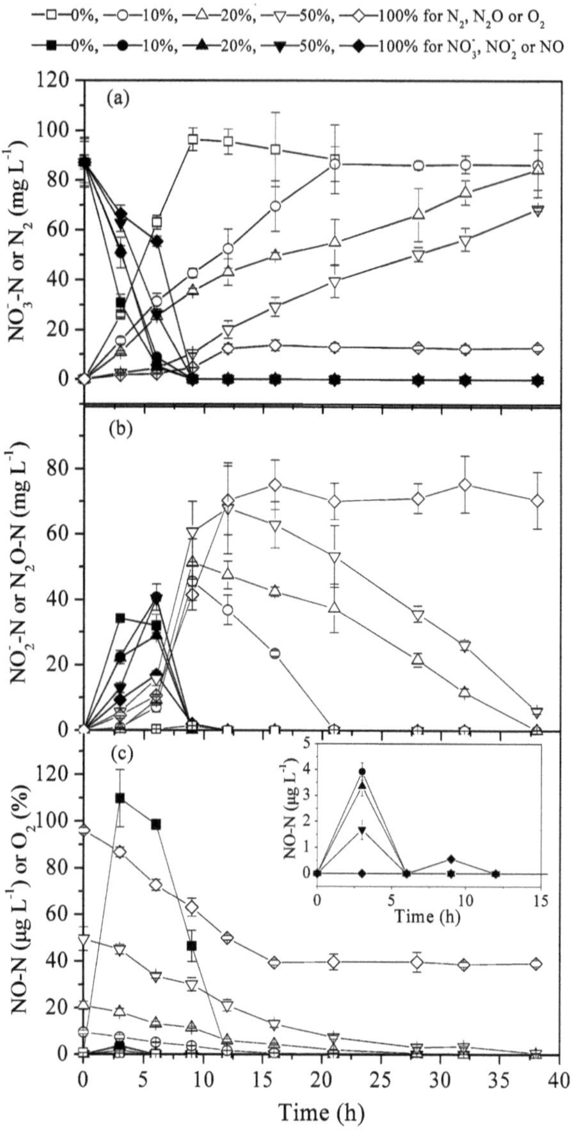

Fig. 5.5 Denitrification performances of strain PCN-2 under different oxygen concentrations. (Reprinted from Zheng et al. (2016), with permission from Elsevier)

under the condition of pure oxygen, and the average nitrate nitrogen degradation rate was 9.69 mg L^{-1} h^{-1}, showing a good ability of oxygen tolerance (Fig. 5.5a).The rate was much higher compared with the previous research results. For example, the degradation rate of nitrate nitrogen was 3.47 mg L^{-1} h^{-1} and 1.03 mg L^{-1} h^{-1} when being decomposed by *P. stutzeri* SU2 and *T. pantotropha* ATCC 35512 under the initial oxygen concentration of 92% (Su et al. 2001). Oxygen was generally considered to be the most competitive electron acceptor in the process of bacterial growth.

Therefore, the existence of oxygen capacity would significantly reduce the denitrifying bacteria (Chen and Strous 2013).

The accumulation of N_2O was not obvious in the experiment, but the N_2O accumulation increased significantly along with the O_2 increased (Fig. 5.5b). The produced N_2O could be transferred to N_2 during the cultured process apart from the pure oxygen conditions. The conversion of N_2O to N_2 stopped after 12 h under the pure oxygen conditions. This might be due to aerobic respiration rate that was too fast, leading to the consumption of the terminal oxidase, which resulted in the lack of energy to maintain the synthesis of N_2O reductase (Bergaust et al. 2008). In addition, under the condition of high oxygen concentration, N_2O reductase activity was seriously inhibited, and bacteria damages by a large number of active free radicals produced in aerobic respiration could also inhibit N_2O reduction (Cabiscol et al. 2000; Otte et al. 1996).

NO showed obvious accumulation under anoxic conditions, and the highest concentration was over 100 μg L^{-1}. With the increase of O_2 concentration, the accumulation of NO decreased significantly (Fig. 5.5c). On the surface, this was not consistent with the general regulation of denitrification by O_2 inhibition, so this phenomenon was likely to be caused by abiotic reactions. In aqueous solution, NO redox activity was very high; it was easy to react with NO_2 and O_2. Studies had shown that the reaction followed first-order kinetics (Kharitonov et al. 1994). Therefore, the removal of NO by chemical pathways might account for the greater proportion in high O_2 concentration conditions. Nevertheless, biological effects might still play an important role in NO removal. First, NO reductase had a hemoglobin center, and its affinity for NO was greater than that of O_2, indicating that it was not at a disadvantage in the electron competition with O_2 (Carr and Ferguson 1990). Second, all copper-type oxidoreductase could restore NO physiologically (Chen and Strous 2013), so the strain PCN-2 had the ability of simultaneous aerobic respiration and NO reduction. In addition, a widely present inner cellular protein called flavohemoglobin globulin could combine with NO and transfer into nitrate under aerobic conditions, thereby reducing the toxicity of NO (Poole and Hughes 2000). Therefore, the chemical reaction and biological action together led to the efficient removal of NO at high O_2 concentration.

5.3.5 *Bioaugmentation of Strain PCN-2 in Activated Sludge*

When the simulated wastewater was treated with nitrate as the major nitrogen source, the denitrification rate of activated sludge system inoculated with strain PCN-2 was significantly increased (Fig. 5.6). No strain PCN-2 was added into the control group, and it required 10 h to degrade 150 mg L^{-1} nitrate completely. However, the process required only 5 h and 2 h when the inoculum size was 15% and 30%, respectively. In addition, the system of 30% strain PCN-2 inoculation did not show obvious nitrite accumulation during denitrification system, with a certain amount of nitrite accumulation appearing at 2 h in 15% inoculated system quickly consumed. Meanwhile,

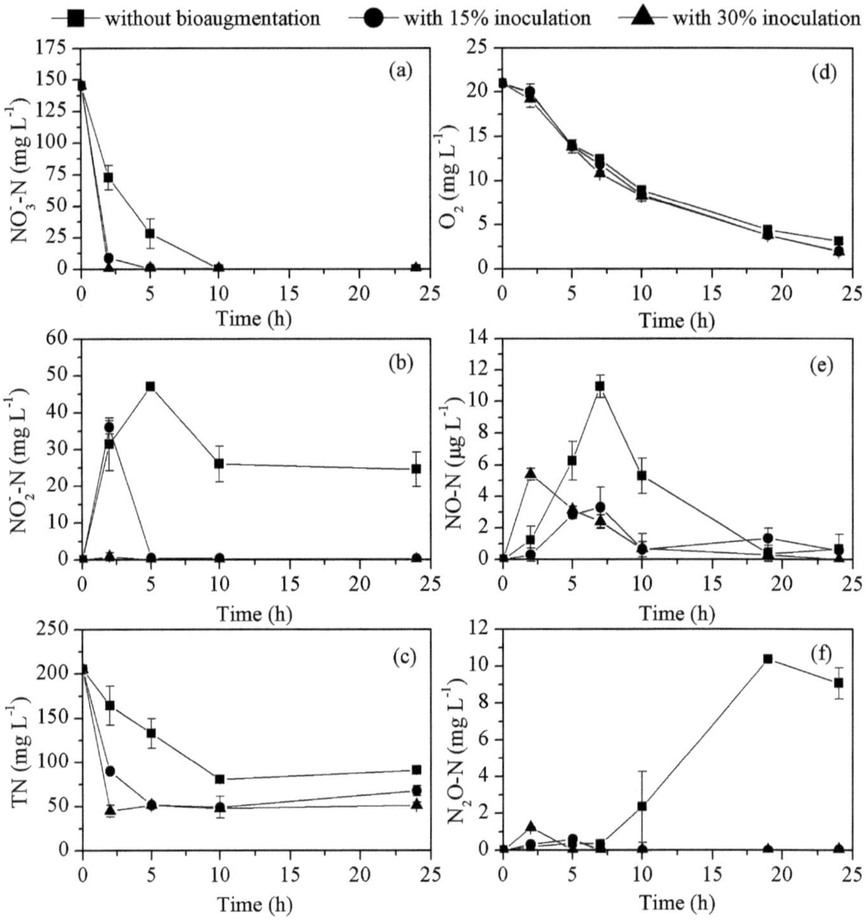

Fig. 5.6 Bioaugmentation of strain PCN-2 in activated sludge when nitrate was the main pollutant

the highest concentration of nitrite nitrogen was 47.06 mg L^{-1} and stabilized at 25 mg L^{-1} after 10 h in the control group. Because the accumulation of nitrite nitrogen is the major factor leading to the NO and N_2O emissions during biological nitrogen removal, a large amount of NO and N_2O was detected in the control group, and the emission of NO and N_2O was greatly reduced when the strain PCN-2 was present.

When a large amount of nitrite nitrogen was present in the simulated wastewater, the rate of nitrite degradation was accelerated obviously by biological enhancement (Fig. 5.7). When the inoculum size was 15% and 30%, the removal rate of 5 h nitrite reached 66.67% and 100%, compared with 46.24% in the control group. From the total nitrogen data, it could be inferred that the removal of nitrite after 5 h was mainly caused by nitrite oxidizing bacteria (NOB) in the activated sludge, which converted

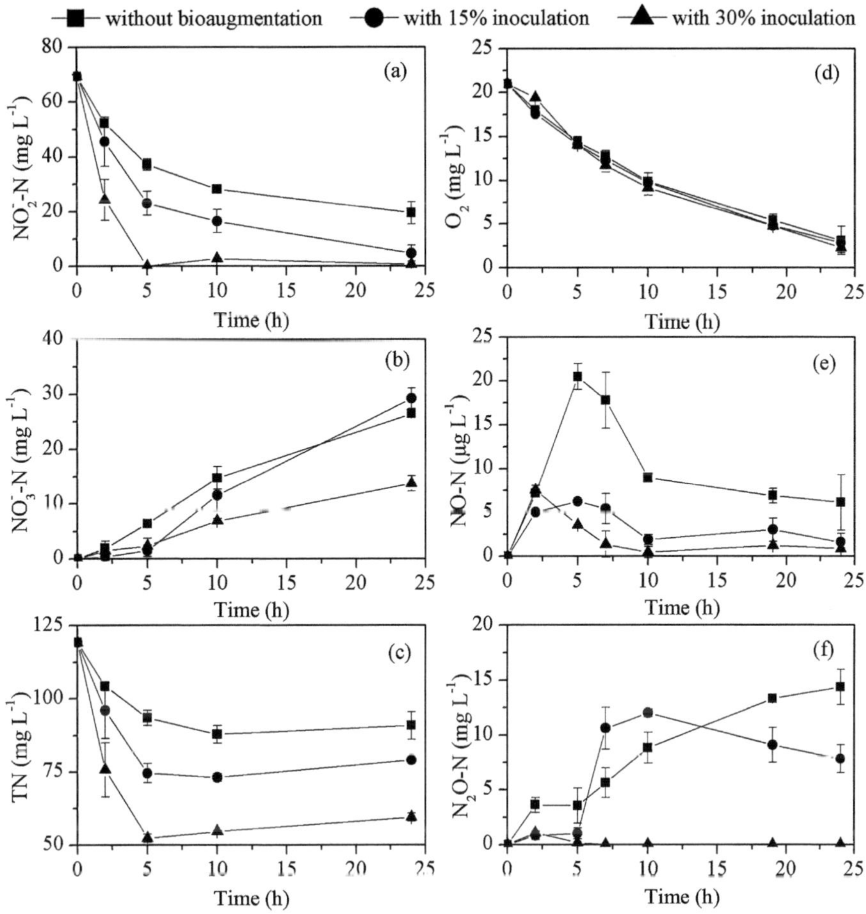

Fig. 5.7 Bioaugmentation of strain PCN-2 in activated sludge when nitrite was the main pollutant

nitrite into nitrate nitrogen. The total nitrogen removal rates of the experimental group were 54.12% and 38.66% at 10 h, which was much higher than 26.05% of the control group. Compared with the experimental results where nitrate nitrogen was the main pollutant, the emission of nitrogen oxides in the experimental group was increased due to high nitrite concentration in the process of nitrogen removal, but in the strain PCN-2-inoculated activated sludge system, NO and N_2O emissions were reduced to some extent, and the emission reduction effect of inoculation was increased, especially in the system of 30% inoculation where no obvious accumulation of N_2O was detected. The experimental results showed that, even under the condition of higher nitrogen oxide production, biological enhancement by inoculation with PCN-2 could significantly reduce the amount of nitrogen oxide emissions from wastewater treatment.

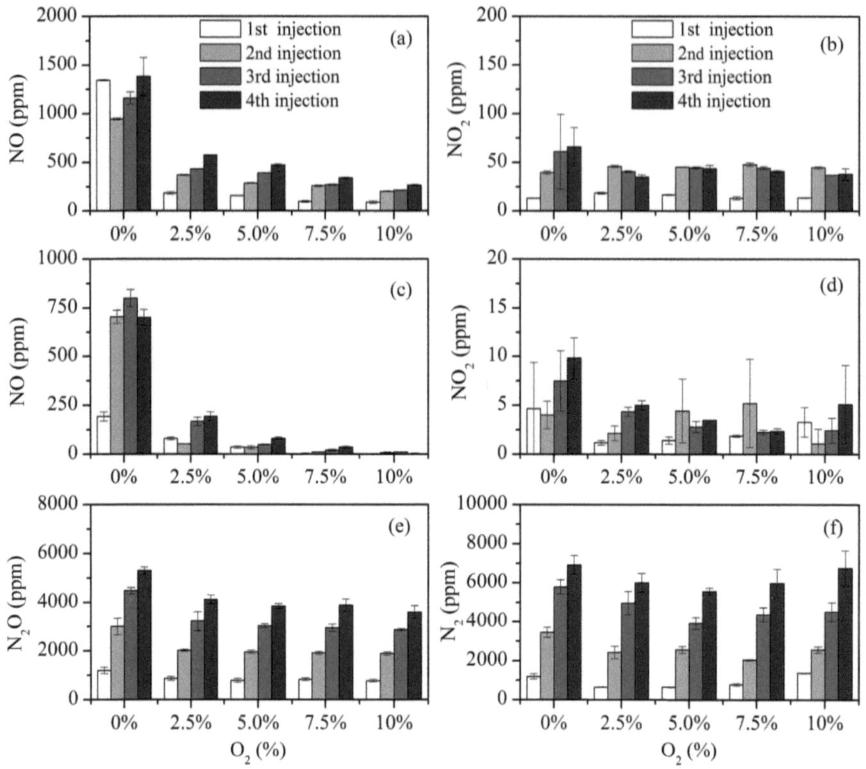

Fig. 5.8 NO and NO_2 removal by immobilized strain PCN-2 in control group (**a** and **b**) and experimental group (**c** and **d**), N_2O and N_2 concentrations in experimental group (**e** and **f**) after each reactions with repetitive NO injections. (Reprinted from Zheng et al. (2016), with permission from Elsevier)

5.3.6 NO_x Removal by Immobilized Strain PCN-2 in Sealed Bottles

As shown in Fig. 5.8, the removal efficiency of NO under five different oxygen concentrations (0%, 2.5%, 5%, 7.5%, and 10%) was investigated experimentally. From the results of the control group, it could be seen that even without bacteria inoculation, NO concentration in 1 h can also significantly decrease with the increase of the residual O_2 concentration (Fig. 5.8a), which was mainly due to rapid chemical reaction between NO, O_2, and water (Formula 5.1 and 5.2). After four times repeated injection of NO, the concentration of nitrate was detected up to 120 mg L^{-1} in the water solution of the sealed bottle, which proved the inference that NO was involved in the chemical reaction.

$$2\,NO + O_2 \rightarrow 2\,NO_2 \quad (5.1)$$

$$3\,NO_2 + H_2O \rightarrow 2\,HNO_3 + NO \quad (5.2)$$

In the sealed bottles of immobilized strain PCN-2, the residual concentration of NO was lower than that of the control group. Therefore, the denitrification of bacteria was also an important cause of NO removal besides chemical reactions (Fig. 5.8c). Similarly, in denitrification, the accumulated NO_2 concentrations in the experimental group were one order of magnitude lower than those in the control group (Figs. 5.8b, d). In the experimental group, the concentration of N_2O and N_2 increased after the injection of NO due to gaseous products by strain denitrification (Fig. 5.8e, f), while this phenomenon was not observed in the control group. With consisting injection of NO, the removal rate of NO decreased in both the control and the experimental group. One of the reasons is that the chemical reaction equilibrium was not conducive to continuous consumption of concentration of NO. On the other hand, the inhibition was due to the consumption of the carbon source in the experimental group and the toxic effects of NO by denitrification activity.

When the NO_2 gas was injected into the sealed bottle, the removal effect of immobilized strain PCN-2 on NO_x was similar to that of NO (Fig. 5.9). In the control group, the concentration of NO_2 decreased greatly due to the chemical reaction, but the accumulation of NO was very high. In the experimental group, NO was denitrified by the strain, and its accumulation amount was one order lower than that of the control group because of the presence of strains. Similarly, denitrification products such as N_2O and N_2 were detected only in the experimental group. The experiment found that the total amount produced higher N_2O and N_2, and the two are not balanced compared with the amount of NO_x decreased. This was because that immobilized strains was cultured for a period of time in a culture medium containing a small amount of nitrogen in order to promote the expression of strain denitrification enzyme. In a static bottle at the beginning of the experiment, a certain quantity of nitrate nitrogen was still contained in immobilized strains, and nitrate might also be used to strain denitrification contributing to total yields of N_2O and N_2. Compared with NO, the accumulated N_2O concentration in the stationary bottle was very low after the injection of NO_2, while the concentration of N_2 accumulation was very high (Fig. 5.9e and f), indicating that the high concentration of NO will hinder the transformation of N_2O to N_2 to a certain extent.

Through the experiments above, we can see that the removal of NO and NO_2 by immobilized strain PCN-2 was prominent, which laid a foundation for its successful application in the trickling filter system.

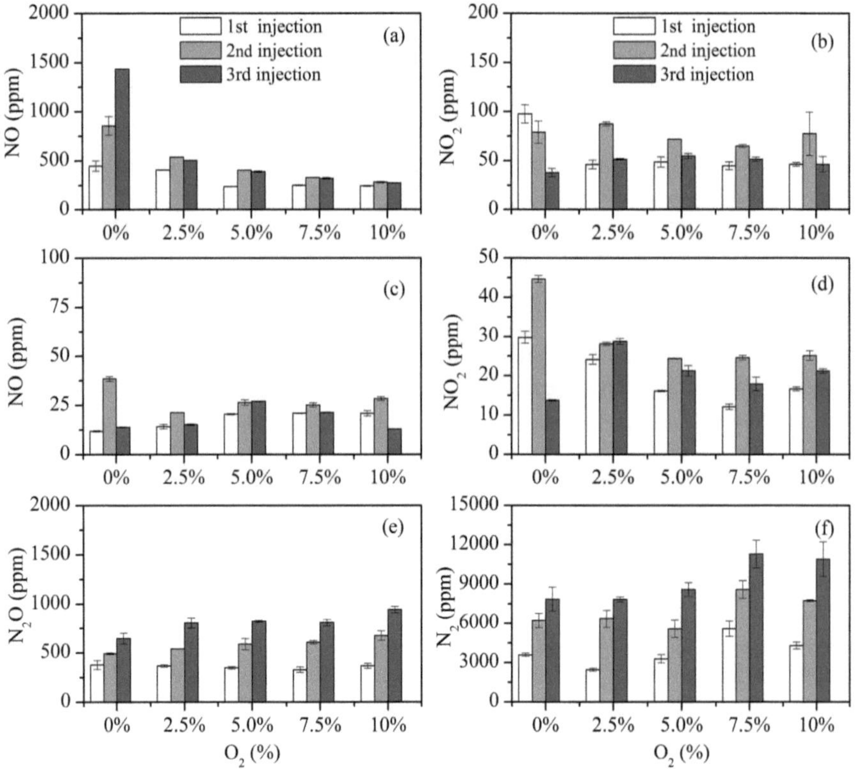

Fig. 5.9 NO and NO_2 removal by immobilized strain PCN-2 in control group (**a** and **b**) and experimental group (**c** and **d**), N_2O and N_2 concentrations in experimental group (**e** and **f**) after each reactions with repetitive NO_2 injections. (Reprinted from Zheng et al. (2016), with permission from Elsevier)

5.3.7 NO_x Removal in Biotrickling Filter with Immobilized Strain PCN-2

In order to investigate the removal efficiency of NO_x in the trickling filter system of immobilized strain PCN-2, the simulated flue gas containing different NO and O_2 concentrations was introduced into the system. The results were shown in Fig. 5.10. Scanning electron microscopy showed that the growth of the biofilm on the surface of the carrier was in good condition after 2 weeks of operation (Fig. 5.11). When the intake NO was increased from 100 ppm to 500 ppm, the NO concentration in the tail gas was increased from 7.6 ppm to 23.2 ppm, but the NOx removal rate was stable in the range of 92.4–95.4% (Fig. 5.10a). Compared with previous studies, the removal rate of NO_x was obviously improved (Jiang et al. 2009; Niu et al. 2014). When the

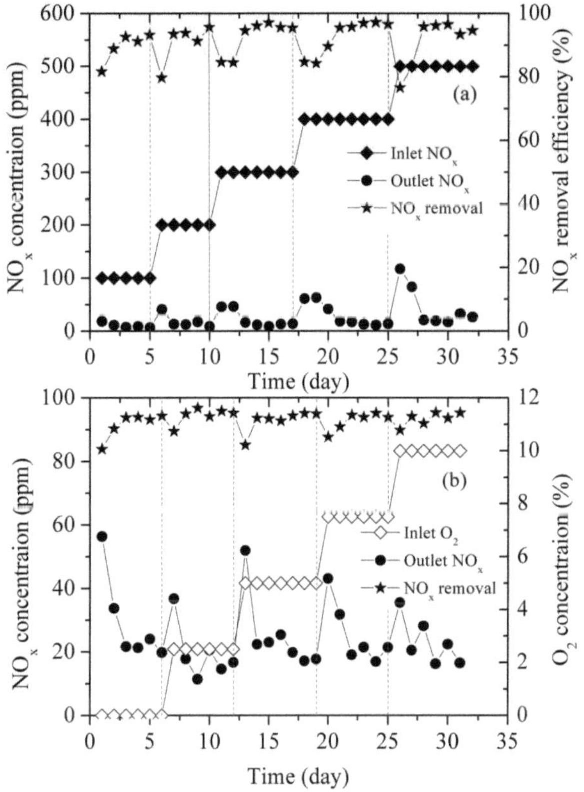

Fig. 5.10 NO_x removal in biotrickling filter under different inlet NO (**a**) and O_2 concentrations (**b**). (Reprinted from Zheng et al. (2016), with permission from Elsevier)

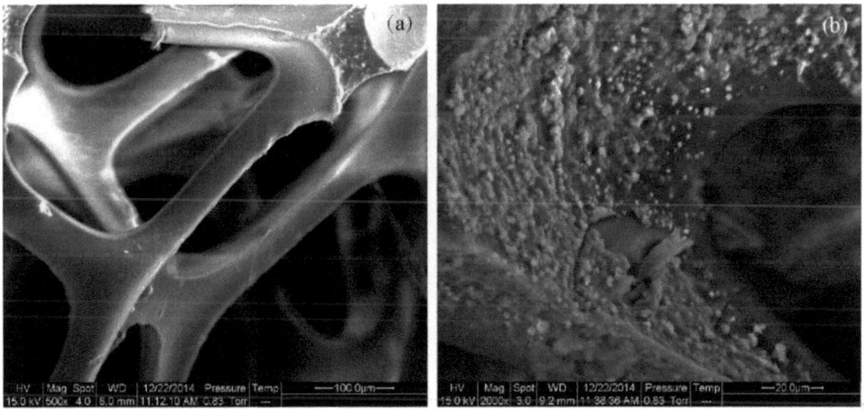

Fig. 5.11 The biofilm of strain PCN-2 formed on the polyurethane foam in biotrickling filter. (Reprinted from (Zheng et al. 2016), with permission from Elsevier)

oxygen concentration in the inlet was increased from 0% to 10%, the NO_x removal rate in the steady state of the system was stable at 90–96% (Fig. 5.10b), indicating that the influence of O_2 on the removal rate of NO_x was not obvious. In addition, when the system was impacted by shock loading of NO or O_2 concentration increased, high removal rates can be restored only after 1D operation despite fluctuations of the removal efficiency of NO_x, indicating that the construction of immobilized strain PCN-2 trickling filter system had a strong resistance and the ability to adapt to environment disturbances rapidly.

In the previous studies, oxygen was usually an important factor in inhibiting NO_x removal, and the inhibition of the system was difficult to restore. The biofilter under hypoxia acclimation was exposed to 2% O_2 after NO removal efficiency dropped significantly. After 15 days NO_x removal began to recover. When exposed to 4% O_2 in biofilter, the NO removal capacity could not be restored (Lee and Apel 1998). Lee et al. (2001) also found that the NO removal rate reached more than 50% under the anoxic conditions. When the biofilter was exposed to 2% O_2, the removal rate of NO was reduced to 10–20% and could not cover back after 55 days. Yang et al. (2007) found that when the intake of O_2 content increased from 0% to 6%, the NO removal rate of the small scale biofilter decreased from 99% to 55%. Therefore, strain PCN-2 showed great potential in flue gas denitrification system.

5.4 Summary

In this chapter, the aerobic denitrification performance of PCN-2 and its application in NO_x removal were studied. The following conclusions were obtained:

1. There was almost no accumulation of NO in aerobic denitrification process of PCN-2 with nitrate or nitrite as the nitrogen source, and N_2 was the final denitrification product. The aerobic respiration and denitrifying enzyme of strain PCN-2 was not inhibited by the high concentration of nitrite.
2. Real-time qPCR analysis showed that the synergistic expression and regulation of denitrifying genes *napA*, *nirS*, and *cnorB* of strain PCN-2 were the main reasons for the high efficiency of NO removal. The high expression of cnorB gene promoted the synthesis of NO reductase and played an important role in suppressing NO accumulation.
3. Strain PCN-2 could exert aerobic denitrification in the range of 0–100% O_2, and the nitrate removal rate was even 9.69 mg L^{-1} h^{-1} under pure oxygen condition. The accumulation of NO was affected by chemical and biological factors and significantly decreased with the increasing of O_2 concentration.
4. In the experiment of bioaugmentation in activated sludge using PCN-2, the nitrogen removal efficiency was enhanced significantly, and the accumulation of NO and N_2O in the denitrification process was significantly reduced whether using nitrate or nitrite as nitrogen source.

5. The results of PCN-2 immobilization in serum bottle experiments showed that PCN-2 had significant effect on NO and NO_2 removal. In the flue gas denitrification system, by immobilizing PCN-2 in a trickling filter reactor, the simulated flue gas NO_x removal rate could reach 90–96% under the oxygen concentration of 0–10% and NO concentration of 100–500 ppm, which has achieved the targets of removing NO_x in flue gas efficiently in the presence of oxygen.

References

Barnes, J. M., Apel, W. A., & Barrett, K. B. (1995). Removal of nitrogen oxides from gas streams using biofiltration. *Journal of Hazardous Materials, 41*, 315–326.

Bergaust, L., Shapleigh, J., Frostegard, A., & Bakken, L. (2008). Transcription and activities of NO_x reductases in agrobacterium tumefaciens: The influence of nitrate, nitrite and oxygen availability. *Environmental Microbiology, 10*, 3070–3081.

Braker, G., & Tiedje, J. M. (2003). Nitric oxide reductase (*norB*) genes from pure cultures and environmental samples. *Applied and Environmental Microbiology, 69*, 3476–3483.

Bru, D., Sarr, A., & Philippot, L. (2007). Relative abundances of proteobacterial membrane-bound and periplasmic nitrate reductases in selected environments. *Applied and Environmental Microbiology, 73*, 5971–5974.

Cabiscol, E., Tamarit, J., & Ros, J. (2000). Oxidative stress in bacteria and protein damage by reactive oxygen spicies. *International Microbiology, 3*, 3–8.

Carr, G. J., & Ferguson, S. J. (1990). Nitric oxide formed by nitrite reductase of Paracoccus denitrificans is sufficiently stable to inhibit cytochrome oxidase activity and is reduced by its reductase under aerobic conditions. *Biochimica et Biophysica Acta, 1017*, 57–62.

Chen, J. W., & Strous, M. (2013). Denitrification and aerobic respiration, hybrid electron transport chains and co-evolution. *Biochimica et Biophysica Acta, 1827*, 136–144.

Cheng, J., Zhu, X., Ni, J., & Borthwick, A. (2010). Palm oil mill effluent treatment using a two-stage microbial fuel cells system integrated with immobilized biological aerated filters. *Bioresource Technology, 101*, 2729–2734.

Jiang, R., Huang, S., Chow, A. T., & Yang, J. (2009). Nitric oxide removal from flue gas with a biotrickling filter using *Pseudomonas putida*. *Journal of Hazardous Materials, 164*, 432–441.

Jin, Y. M., Veiga, M. C., & Kennes, C. (2005). Bioprocesses for the removal of nitrogen oxides from polluted air. *Journal of Chemical Technology and Biotechnology, 80*, 483.

Kester, R. A., De Boer, W., & Laanbroek, H. J. (1997). Production of NO and N_2O by pure cultures of nitrifying and denitrifying bacteria during changes in aeration. *Applied and Environmental Microbiology, 63*, 3872–3877.

Kharitonov, V. G., Sundquist, A. R., & Sharma, V. S. (1994). Kinetics of nitric oxide autoxidation in aqueous solution. *Journal of Biological Chemistry, 269*, 5881–5883.

Knowles, R. (1982). Denitrification. *Microbiology Reviews, 46*, 43–70.

Lee, B. D., & Apel, W. A. (1998, October). *Evaluation of NO_x removal using compost based biofilters operated under denitrifying conditions.* (pp. 51–60). Proceedings of the 1998 USC-TRG Conference on Biofiltration, Los Angeles.

Lee, B. D., Apel, W. A., & Smith, W. A. (2001). Oxygen effects on thermophilic microbial populations in biofilters treating nitric oxide containing off-gas streams. *Environmental Progress, 20*, 157–166.

Lu, Y., Wang, J., Yu, Y., Shi, L., & Kong, F. (2014). Changes in the physiology and gene expression of Microcystis aeruginosa under EGCG stress. *Chemosphere, 117*, 164–169.

Muyzer, G., De Waal, E. C., & Uitterlinden, A. G. (1993). Profiling of complex microbial populations by denaturing gradient gel electrophoresis analysis of polymerase chain reaction-amplified genes coding for 16S rRNA. *Applied and Environmental Microbiology, 59*, 695–700.

Niu, H., Leung, D. Y., Wong, C., Zhang, T., Chan, M., & Leung, F. C. (2014). Nitric oxide removal by wastewater bacteria in a biotrickling filter. *Journal of Environmental Sciences-China, 26*, 555–565.

Otte, S., Grobben, N. G., Robertson, L. A., Jetten, M. S., & Kuenen, J. G. (1996). Nitrous oxide production by *Alcaligenes faecalis* under transient and dynamic aerobic and anaerobic conditions. *Applied and Environmental Microbiology, 62*, 2421–2426.

Philippot, L. (2002). Denitrifying genes in bacterial and archaeal genomes. *Biochimica et Biophysica Acta, 1577*, 355–376.

Poole, R. K., & Hughes, M. N. (2000). New functions for the ancient globin family: Bacterial responses to nitric oxide and nitrosative stress. *Molecular Microbiology, 36*, 775–783.

Robertson, L. A., Dalsgaar, T., Revsbech, N. P., & Kuenen, J. G. (1995). Confirmation of 'aerobic denitrification' in batch cultures, using gas chromatography and ^{15}N mass spectrometry. *FEMS Microbiology Ecology, 18*, 113–120.

Saleh-Lakha, S., Shannon, K. E., Goyer, C., Trevors, J. T., Zebarth, B. J., & Burton, D. L. (2008). Nitric oxide reductase gene expression and nitrous oxide production in nitrate-grown *Pseudomonas mandelii. Applied and Environmental Microbiology, 74*, 6876–6879.

Spiro, S. (2012). Nitrous oxide production and consumption: Regulation of gene expression by gas-sensitive transcription factors. *Philosophical Transactions of the Royal Society B, 367*, 1213–1225.

Su, J. J., Liu, B. Y., & Liu, C. Y. (2001). Comparison of aerobic denitrification under high oxygen atmosphere by *Thiosphaera pantotropha* ATCC 35512 and *Pseudomonas stutzeri* SU2 newly isolated from the activated sludge of a piggery wastewater treatment system. *Journal of Applied Microbiology, 90*, 457–462.

Throback, I. N., Enwall, K., Jarvis, A., & Hallin, S. (2004). Reassessing PCR primers targeting *nirS*, *nirK* and *nosZ* genes for community surveys of denitrifying bacteria with DGGE. *FEMS Microbiology Ecology, 49*, 401–417.

Vollack, K. U., & Zumft, W. G. (2001). Nitric oxide signaling and transcriptional control of denitrification genes in *Pseudomonas stutzeri. Journal of Bacteriology, 183*, 2516–2526.

Wan, C. L., Yang, X., Lee, D. J., Du, M. A., Wan, F., & Chen, C. (2011). Aerobic denitrification by novel isolated strain using NO_2^--N as nitrogen source. *Bioresource Technology, 102*, 7244–7248.

Woertz, J. R., van Heiningen, W. N., van Eekert, M. H., Kraakman, N. J., Kinney, K. A., & van Groenestijn, J. W. (2002). Dynamic bioreactor operation: Effects of packing material and mite predation on toluene removal from off-gas. *Applied Microbiology and Biotechnology, 58*, 690–694.

Yang, W. F., Hsing, H. J., Yang, Y. C., & Shyng, J. Y. (2007). The effects of selected parameters on the nitric oxide removal by biofilter. *Journal of Hazardous Materials, 148*, 653–659.

Zheng, M., Li, C., Liu, S., Gui, M., & Ni, J. (2016). Potential application of aerobic denitrifying bacterium Pseudomonas aeruginosa PCN-2 in nitrogen oxides (NOx) removal from flue gas. *Journal of Hazardous Materials, 318*, 571–578.

Chapter 6
Conclusions and Outlook

6.1 Conclusions

In this paper, two highly efficient aerobic denitrifying bacterial strains that can quickly utilize nitrogen oxides were selected aiming to solve the tough problem of large amount of nitrogen oxide emission from biological nitrogen removal process. Based on the deep insight into the characteristics and mechanism of nitrogen oxide emission reduction, the strains were applied to activated sludge- and biofilm-based reactors, which have achieved significant bioaugmentation in improving denitrification efficiency and nitrogen oxide abatement. The main conclusions are as follows:

1. Two bacterial strains capable of rapid utilization of nitrogen oxide were selected from 14 aerobic denitrifying bacterial strains, which were identified as *Pseudomonas stutzeri* PCN-1 and *Pseudomonas aeruginosa* PCN-2, respectively.
2. The removal rates of nitrate, nitrite, and N_2O nitrogen by *P. stutzeri* strain PCN-1 were 11.66 mg L^{-1} h^{-1}, 12.80 mg L^{-1} h^{-1}, and 9.67 mg L^{-1} h^{-1} under aerobic condition. Meanwhile, the maximum accumulation amount of NO and N_2O were only 0.003% and 0.33% of the total removed nitrate and 0.006% and 0.29% of the total removed nitrite. The co-expression of denitrifying functional genes *nirS*, *cnorB*, and *nosZ* of strain PCN-1 was the main cause for its capability of aerobic denitrification. Strain PCN-1 had the strongest denitrification ability and the lowest accumulation of NO and N_2O under the condition of pH of 7–9 and C/N of 4–6 with sodium succinate or sodium acetate as carbon source. Nitrate could be completely degraded by strain PCN-1 under O_2 concentration ranging from zero to 100%. NO concentration decreased significantly with the increase of O_2 concentration, and the accumulation of N_2O significantly increased with the increase of O_2 concentration. However, N_2O accumulation only accounted for

M. Zheng, *Nitrogen Removal Characteristics of Aerobic Denitrifying Bacteria and Their Applications in Nitrogen Oxides Emission Mitigation*, Springer Theses,
https://doi.org/10.1007/978-981-13-2432-1_6

1.12% of the total nitrate removal under O_2 concentration of 50%, demonstrating a strong N_2O reduction ability with resistance to high oxygen concentration.
3. Nitrogen oxide emission was significantly reduced accompanied with enhanced aerobic nitrogen removal when strain PCN-1 was applied to the SBR systems. Under the oxygen content of 0%, 5%, and 10% in the aeration, the N_2O emission of SBR decreased by 19.81%, 46.12%, and 45.13% after bioaugmentation, respectively. In a pilot Carrousel oxidation ditch, the N_2O emission rate and dissolved N_2O concentration showed a unique spatial distribution. The contribution of N_2O emission from aeration zone to the total emission of N_2O in oxidation ditch reached 91.14%. The contribution of heterotrophic denitrification and autotrophic denitrification to the total amount of N_2O in the whole oxidation ditch system was 10.22% and 89.78%, respectively. When the oxidation ditch system was steadily operated at the sludge age of 17 days, N_2O and NO emissions from the aerobic zone decreased by 36.6% and 23% with the bioaugmentation of strain PCN-1. When the system was subjected to high ammonia load, N_2O and NO emissions were reduced by 49.13% and 61.87%; when subjected to aeration failure, N_2O and NO emissions were reduced by 59.7% and 64%, clearly demonstrating that strain PCN-1 could significantly improve the stable operation ability of the activated sludge system and mitigate nitrogen oxide emissions caused by shock loading.
4. Almost no NO was accumulated in denitrification of strain *P. aeruginosa* PCN-2 with nitrate or nitrite as nitrogen source. The co-expression of denitrification genes *napA*, *nirS*, and *cnorB* was the main reason for the efficient NO removal. Strain PCN-2 could play normal denitrification function in the range of O_2 concentration of 0–100%, and the amount of NO emission decreased with the increase of O_2 concentration. Shake flask test showed that strain PCN-2 could significantly enhance the denitrification efficiency and reduce the emission of NO and N_2O of the activated sludge system. The experiment of stationary sealed bottle showed that the immobilized strain PCN-2 had prominent removal capability of NO and NO_2. Finally, the immobilized strain PCN-2 was applied in trickling filter reactor to construct the flue gas purification system, in which NO_x removal rate reached 90–96% under oxygen concentration of 0–10% and NO concentration of 100–500 ppm, displaying a great application potential in nitrogen oxide emission reduction.

6.2 Outlook

In the future, research work can be carried out from the following aspects:

1. The genomes of the two high-efficient strains should be sequenced, and then the specific nitrogen metabolism pathway might be discovered through comparing with the standard bacterial strain.

2. In view of the affinity differences of the two strains to the substrate NO and N_2O, add both of them with a certain proportion in biological enhancement so as to achieve synergistic effects and achieve better effects of NO_x emission reduction.
3. After bioaugmentation, the survival status of the strains can be monitored by molecular biological methods. Meanwhile, comprehensively analyze the microbial structure with high-throughput sequencing to elucidate the significant effect of bioaugmentation on the improvement of the community structure.

Appendix

Appendix S1 Standard Curve of O_2, N_2O and $^{15}N_2$

Fig. S1 Standard curve of O_2, N_2O and $^{15}N_2$ determined by Gas chromatography

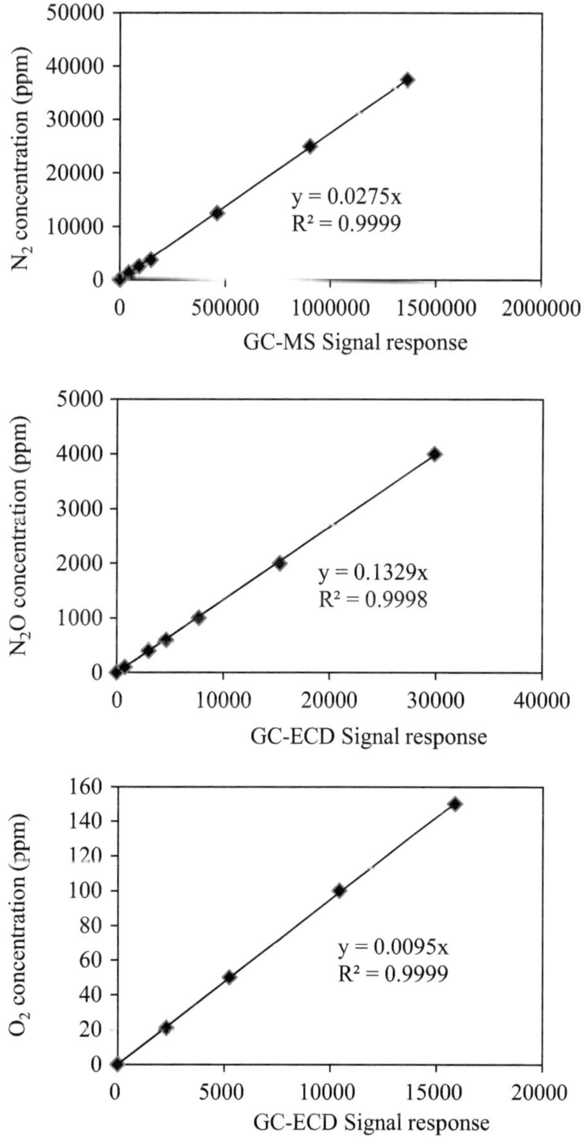

M. Zheng, *Nitrogen Removal Characteristics of Aerobic Denitrifying Bacteria and Their Applications in Nitrogen Oxides Emission Mitigation*, Springer Theses,
https://doi.org/10.1007/978-981-13-2432-1

Appendix S2 Standard Curve of Dry Weight for the Two Strains

Fig. S2 Standard curve of dry weight versus OD_{600} for strain PCN-1 (**a**) and PCN-2 (**b**)

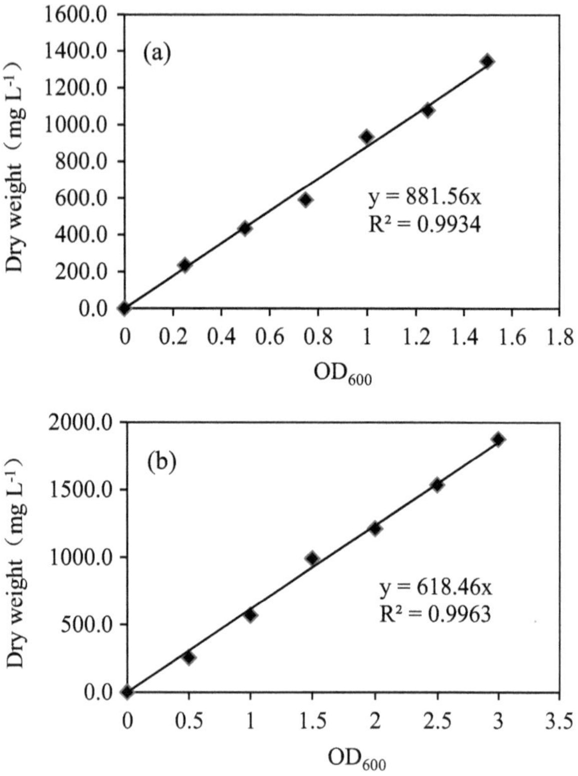

MIX
Papier aus verantwortungsvollen Quellen
Paper from responsible sources
FSC® C105338
FSC
www.fsc.org

If you have any concerns about our products,
you can contact us on
ProductSafety@springernature.com

In case Publisher is established outside the EU,
the EU authorized representative is:
Springer Nature Customer Service Center GmbH
Europaplatz 3, 69115 Heidelberg, Germany

Printed by Libri Plureos GmbH
in Hamburg, Germany